LABORATORY EXERCISES IN MICROBIOLOGY

D0002373

Robert A. Pollack
Nassau Community College

Lorraine Findlay, PHD
Nassau Community College

Walter Mondschein, PHD
Nassau Community College

R. Ronald Modesto
C.W. Post Campus of Long Island University

JOHN WILEY & SONS, INC.

ACQUISITIONS EDITOR	Bonnie Roesch
MARKETING MANAGER	Clay Stone
SENIOR PRODUCTION EDITOR	Kelly Tavares
SENIOR DESIGNER	Karin Kincheloe
SENIOR ILLUSTRATION EDITOR	Anna Melhorn
PHOTO EDITOR	Sara Wight

This book was set in 10.5/13 Times Roman by TechBooks and printed and bound by Hamilton Press. The cover was printed by Lehigh Press.

ISBN 0-471-41412-3

Printed in the United States of America

10 9 8 7 6 5 4 3 2 1

PREFACE

Developed for use in an undergraduate microbiology laboratory course, *Laboratory Exercises in Microbiology* meets the needs of students majoring in diverse programs such as allied health or biological sciences. The manual contains a variety of interactive activities and experiments that teach students the basic concepts of microbiology and support the content covered during lectures.

APPROACH AND ORGANIZATION

We are firmly committed to the idea that a microbiology laboratory — and the manual used for it — should extend the learning experience for students, and not be a repeat or reproduction of lecture material. With this in mind we made every effort to avoid duplicating text and illustrations that will be found in their lecture text. We have minimized the amount of textual material for students to read at the start of the laboratory period. Rather, labs are introduced in a clear and concise manner and maintain a student friendly tone. This leaves plenty of time for students to engage in activities and experiments that promote a deeper understanding of microbiological concepts and principles, to answer questions and to write up lab reports.

The 25 Exercises are divided into five Parts: General Microscopy and Aseptic Technique; Microbial Morphology, Differential Stains; Microbial Control and Biochemistry; Medical Microbiology; and Food and Environmental Microbiology. A Photographic Atlas of sixty-eight full-color plates depicting laboratory techniques and results, and numerous micrographs is included.

LABORATORY FEATURES

The self-contained laboratory exercises in this manual are all designed to maximize the learning opportunity and time spent during each laboratory period. Each Exercise begins with a clearly defined list of *Objectives*, followed by a *Materials List* needed to complete the lab. *Procedures* and *Results* sections within each lab are easily identified. Included within lab exercises are a variety of activities like *Art Labeling, Coloring, Identification Exercises* and *Critical Thinking Questions* that help summarize the concepts covered in the lab. An inventory list that allows students to double check their work is also a feature of most exercises.

Immediately following each Exercise is a **Laboratory Report** that can be completed and turned in to an instructor. Included in the lab reports are a mix of *Fill-in-the-blank, Matching* and *Multiple Choice Questions*. At the end is a glossary of **Working Definitions and Terms** that provide a quick review and reinforcement for students as they complete the reports.

SUPPORTING MATERIALS

A Companion Website for instructors complements the use of this manual. It includes the following:

- Laboratory Materials List and Suggestions
- Reagents and Stain Formulations
- Media Formulations
- Suggested Sequencing of laboratories for Health Science students and for non-Health Science students.

- An Answer Key to the questions from the laboratory reports.

The supporting website can be accessed at www.wiley.com/college/pollack

ACKNOWLEDGEMENTS

We wish to acknowledge the assistance provided by Dr. Aleta Labiento, Ph.D in Infectious Disease Control, Walden University, in developing the Laboratory Operation and Safety Instructions located at the beginning of the manual, as well as Exercise One. Dr. Labiento is currently an Associate Professor at York College, New York.

We would also like to recognize the publishing team at Wiley who supported and guided our efforts to produce this manual. Ronde Bradley, our Wiley sales representative, introduced us — and promoted our work — to the editorial team lead by Bonnie Roesch, Senior Editor. Mary O'Sullivan, Associate Editor, worked closely with us to effectively manage the development and review process. Her day to day efforts helped tremendously in fine tuning the final manuscript for this manual. Kelly Tavares managed the production process with skill and care. Hilary Newman coordinated the photo research while Anna Melhorn directed the creation of the illustrations included throughout. Karin Kincheloe created the effective design for both the manual and the cover. Our thanks go to all for their expertise and collaboration.

Finally, we wish to thank the following reviewers who gave us such helpful feedback during the development process:

William H. Coleman, University of Hartford
Teresa G. Fischer, Indian River Community College
Van H. Grosse, Columbus State University
James B. Herrick, James Madison University
Hinrich Kaiser, La Sierra University
Michael A. Lawson, Missouri Southern State
Paul McLaughlin, Madisonville Community College
Barbara Moore, University of Texas
Clarence C. Wolfe, Northern Virginia Community College

We invite all readers and users of this manual to send any comments and suggestions to us so that we can include them in planning for future editions.

Robert A. Pollack
Lorraine Findlay, PHD
Walter Mondschein
Nassau Community College
R. Ronald Modesto, PHD
C.W. Post Campus of Long Island University

CONTENTS

Introduction: Laboratory Operations and Safety

Safety is an important consideration in any laboratory environment. In microbiology, we have the additional concern that comes from utilizing dangerous or potentially dangerous organisms called **pathogens.** The following section lists safety rules appropriate for any laboratory. The procedures and techniques you will learn here will continue to be useful to you in other laboratory courses, at home, and in the workplace for years to come. Please review and familiarize yourself with these procedures so that your laboratory experience will be an enjoyable and SAFE one. Additional safety procedures and requirements specific to microbiology will be reintroduced and reinforced in Exercise 1 and, where appropriate, throughout the manual.

GENERAL LABORATORY OPERATING PROCEDURES

1. *Be prompt.* Microbiology laboratories require that you master various techniques needed to handle and manipulate microbes safely and efficiently. Instruction and demonstrations of these procedures will be done at the beginning of each session.

2. *Be prepared.* If instructed to do so, read the introductory material ahead of time so that you will know what to expect and what is expected of you. Make sure you have your laboratory manual, lab coat, marking pen, and whatever else is required of you for each session.

3. *Be responsible.* Take care of your work area and equipment assigned to you. Wipe down your work area with disinfectant solution before and after each laboratory session. Keep your microscope clean and in good working order. Follow your laboratory in-structor's direction in cleaning, setting up, and putting away your microscope. Be aware of the proper containers needed to place used slides, tubes, stains, chemicals, paper, and other items used in the lab. Leave the laboratory area in good order and return all materials and equipment to their original location.

LABORATORY SAFETY

1. Never put anything up to your mouth in the laboratory. A major way microbes enter the body (called a portal of entry) is by the mouth. Therefore, no eating, drinking, gum chewing, application of cosmetics, or smoking is allowed in the laboratory.

2. Wear a laboratory coat or apron in class. This item of equipment will protect you from various stains, chemicals, and microbes (including those in aerosols). When leaving the lab, either leave the lab coat in an assigned cabinet or place it in a plastic bag or container. This item of apparel is not for street use, regardless of what you may see on TV medical shows.

3. Wash your hands after completing the laboratory session. Do not bring anything up to your mouth or eyes without washing your hands first.

4. Be aware of the locations of the fire extinguisher, eyewash station, deluge shower, and exits.

5. Inform your instructor about any accidents, spills, or potential hazards.

6. When in doubt, ask your instructor about a procedure.

7. Do not wear sandals or open-toed shoes.

8. Do not apply makeup.

9. Tie back long hair.

HANDWASHING

Handwashing is one of the most important procedures used to prevent the spread of microbes from one area to another. Even plain soap can effectively remove significant numbers of microbes from a work surface or a person, thus lowering the chances of infection. You do not have to perform the extremely thorough five minutes of scrubbing associated with operating room procedures. If you followed such a rigorous procedure in your everyday routine, you would not have too much in the way of skin by the end of the day.

Proper handwashing involves scrubbing the hands with soap and water for at least 30 seconds. The soap loosens and sometimes kills the microbes, while the friction due to scrubbing removes them. Specifically, you should practice the following routine after each laboratory session:

1. Remove any rings and bracelets, storing them in a safe location during the lab session. Then place some liquid soap in the palm of your hand. (Bar soap is virtually never used in the clinical area because it may act as a source of microbial contamination.)

2. Lather the soap up using friction to loosen and remove dirt, dead skin, or other contaminants. Pay particular attention to the areas between the fingers and the fingernails. Using the thumb of one hand, rub the cuticle and nail bed of each finger of the opposite hand. If the area under the nails is dirty, use a nail brush.

3. Rinse the soap from your hands; hold the hands in a downward position.

4. Repeat steps 1–3 if your hands are excessively dirty or if you suspect contamination with blood or blood products.

5. Wipe your hands with a paper towel. While holding the paper towel in your hand, turn off the water; otherwise, you will recontaminate your hand.

Always wash your hands:

- After each laboratory session.
- After each patient/client contact.
- If contaminated with any potential infectious material, such as blood, other body fluids, excretions, secretions, or microbial cultures.
- After any procedure in which you have to wear protective gloves.

I GENERAL MICROSCOPY AND ASEPTIC TECHNIQUE

Over 300 years ago, a Dutch merchant, Anton van Leeuwenhoek* (1632–1723), placed a drop of water on a platform and observed it with a home-made lens made of high-quality ground glass. The "animicules" he observed were previously unknown to the scientific community, and so the science of microbiology was born. Leeuwenhoek was the first to observe fungi, protozoa, sperm cells, and even bacteria (at least the larger ones). Modern microscopes, with numerous improvements and much better magnification, have become the present-day microbiologist's mainstay in exploring this previously invisible world.

*Anton van Leeuwenhoek was one of the first modern scientists, although he had no training in the field. Indeed, at the time, no one had any training as a scientist. In his earlier years, he was a cloth merchant and he probably started to grind the lenses he eventually used in his microscopes to inspect the weave of the cloth he bought. Later in life, he was actually the custodian of the local town hall. All during this time, he carefully observed everything he could fit under his microscopes and sent news of his observations to the early scientists of the time.

Since Leeuwenhoek's time, some of these microbes have been identified as the cause of diseases, others are used in the treatment of diseases, and still others have been found to actually help prevent disease. In addition, some microbes have been the source of food for humans, and, conversely, the cause of famines for humans. Many microbes have even changed the ecology of major parts of this planet.

In this manual, you will be introduced to this new microbial world by learning how to use the "tools of the trade," that is, the microscope, so that you can observe and study cells many times smaller than the typical human cell. You will also be introduced to methods that will allow you to move, or transfer, these microbial cells from one type of growth environment to another *safely* and efficiently. By mastering these techniques, you will be able to handle all types of microbes with confidence, even the more dangerous ones called pathogens. The procedures you will learn will also carry over into other parts of the medical field such as dealing with patients infected with these pathogens, without contaminating yourself, your work area, or your family.

Laboratory Safety: Introduction to the Microscope

Objectives

After completing this lab, you should be able to:

1. Explain why no eating, drinking, gum chewing, application of lipstick or other makeup, or smoking is permitted in the laboratory.

2. Locate the fire extinguishers and emergency exits in the laboratory.

3. Locate the fire blanket, eyewash station, and emergency shower.

4. Describe the procedure to follow when a chemical spill or other laboratory accident occurs. Identify the biohazard waste disposal container and the sharps container, and explain their correct usage.

5. Follow the proper procedures for starting and completing each laboratory session.

6. With your laboratory manual as a guide, focus your assigned microscope properly using the low-power, high-dry, and oil immersion objectives.

7. Be familiar with the most common mistakes students make in using the microscope.

8. Prepare a simple stain and recognize the differences between eukaryotic and prokaryotic cells.

Materials (provided by the student if required)

Laboratory manual assigned for each session

Lab coat with long sleeves and long enough to cover the lap when sitting

Permanent marking pen

Safety eyewear

LABORATORY SAFETY AND PROCEDURES

One of the most important aspects of working in a microbiology laboratory is learning, and then following, established procedures for safety—*your safety*—as well as that of others. Besides awareness of fire and chemical hazards, you must also understand that many microbes, if mishandled, are potentially hazardous to humans.

⚠ **SAFETY RULE:** NEVER PUT ANYTHING IN YOUR MOUTH WHILE IN THE LABORATORY, INCLUDING TOUCHING YOUR MOUTH WITH YOUR HANDS, TIPS OF PENS OR PENCILS, OR MOUTH PIPETTES.

⚠ **SAFETY RULE:** DO NOT PLACE BACKPACKS, POCKETBOOKS, OR COATS ON THE LAB TABLES.

⚠ **SAFETY RULE:** AFTER EACH LAB IS FINISHED, WASH YOUR HANDS IMMEDIATELY WITH DISINFECTANT SOAP.

In this course, you will assume that *all* microbes are dangerous. One of the major ways that these microbes enter the body is by way of the mouth; thus, let us reiterate: no eating, drinking, gum chewing, or smoking in the lab, or anywhere else where hazardous microbes are routinely encountered.

Before and after each laboratory session, you will disinfect the tabletops and wash your hands immediately.

⚠ **SAFETY RULE:** REPORT ALL CUTS OR WOUNDS TO YOUR INSTRUCTOR.

Another way microbes can enter your body is through damaged skin. If you have any cuts or scrapes on your hand when entering the lab, or if you receive a wound that damages the skin while in the lab, make sure it is covered properly with an appropriate bandage or gloves before you perform any laboratory procedure.

In addition, you should observe the location(s) of various first aid and items of safety equipment around the laboratory. Listen carefully to your instructor's directions on how to use the fire blanket, eyewash station, and deluge showers and other safety equipment specifically utilized in your lab. Note the location of the fire extinguisher and exits. In the unlikely event of a laboratory fire, exit the building with your lab instructor. **DO NOT** leave your group for any reason until the Fire Marshal, Fire Department, or Security gives you permission to do so.

⚠ **SAFETY RULE:** REPORT ALL SPILLS AND ACCIDENTS TO YOUR INSTRUCTOR.

Inhalation is also a means by which dangerous substances enter the body.

⚠ **SAFETY RULE:** NEVER ENTER OR REENTER A LAB ROOM WHERE THERE IS SMOKE. TOXIC FUMES OFTEN DO NOT HAVE A STRONG ODOR.

During the remaining laboratory sessions, you will be working with many different kinds of microbes and chemicals. Although most of the materials with which you come into contact will not pose any real danger, we will assume that they are all potentially hazardous and must be handled appropriately. When you are required to use certain toxic chemicals, such as Kovac's reagent, your instructor will give you additional guidelines to follow.

⚠ **SAFETY RULE:** NEVER LEAVE THE LABORATORY WITH YOUR LAB COAT ON. PLACE YOUR COAT IN A POLY BAG BEFORE YOU LEAVE. REMOVE IT FROM THE LABORATORY ONLY TO WASH IT.

LABORATORY PROCEDURES

Microbiology laboratory procedures have long been established to keep bacteria, fungi, viruses, and other types of microbes under control while they are being studied. Such procedures have helped enhance the life span of many medical workers over the last 150 years. Cleanliness is an important aspect of this control as well as the various activities that have generally became known as *aseptic technique* (the word aseptic deriving from Sepsis = infection or infectious material, A = without). One aspect of aseptic technique involves working with infectious material without getting infected. More sophisticated aspects of aseptic technique will be covered later in this course as more information about the field of microbiology becomes available to you.

To explain all of the aseptic technique procedures at this time would be somewhat overwhelming, for it is important to know both *why* you are performing a specific action and *how* to perform that action. This laboratory manual therefore addresses the preparation of the work area before actual microbiology activities take place as well as the procedures needed to clean up after the lab exercise is completed. As part of future lab exercises, other aseptic techniques will be introduced so that your knowledge of such procedures will gradually increase over the course of the semester.

The first and the last laboratory procedure you will perform every day is the cleaning of your work area with a disinfectant solution. Some disinfectants also provide the added benefit of serving as a cleaning agent. By wiping down your work area before the lab starts, you kill or remove microbes that may contaminate your work, as well as remove any residual dye, stain, or oil left by previous classes. And by repeating the same task at the end of class, you are leaving a clean, safe area for the next person.

SIMPLE STAIN TECHNIQUE

Stains provide better visualization of most objects seen under the microscope. Without such stains, cells are nearly transparent and extremely difficult to see. A *simple stain* is one in which a single colored dye or stain is used to tint the object. If a red dye is used as the stain, everything that accepts or absorbs the stain will appear red under the microscope. With certain types of cells, such as human cells, different components or organelles absorb different amounts of dye; thus, you will see various shades of the same color within the same cell. Because the nucleus of the human cell has a greater affinity for most dyes than the cytoplasm, the nucleus tends to show a darker shade of color than the rest of the cell.

Materials/Student

Toothpick or cotton swab

Glass slides

China marking pencil or permanent marker

Staining tray

Bibulous paper or paper towel

Methylene blue stain or crystal violet stain

Prepared slides of cocci, bacilli, and spirilla

PROCEDURE

1. Prepare a cheek smear by *gently* scraping a toothpick or handle of a cotton swab along the inside of your cheek and then smearing this material on a marked section of a glass slide. *Discard* the toothpick or cotton swab in a disinfectant solution or as directed.

   (*Note:* This is an example of aseptic technique. Any sample taken from an individual is considered potentially hazardous and cannot be thrown away in a regular garbage pail or wastepaper basket.) Figure 1.1 shows the preparation of a smear.

2. Allow the smear to *air dry* (letting the moisture evaporate until the slide is dry) and then *heat fix* (holding the bottom of the slide in the flame of a Bunsen burner for a second or so) as directed. You may be directed to use a clothes pin on the slide to prevent possible singed fingers. This procedure will adhere proteins, and thus cells, on the slide, as well as function to kill many cells.

3. Add enough *crystal violet* stain to cover the smear, and leave it on for approximately 5 seconds. If you are using *methylene blue*, leave the stain on for 1 minute. Your instructor will tell you which one to use. Figure 1.2 shows the stain being added to the slide on a staining tray.

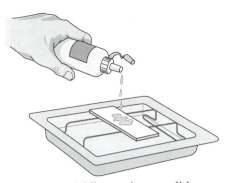

FIG. 1.2. Adding stain to a slide.

FIG. 1.3. Rinsing off a slide.

4. Rinse off with tap water from the sink or with the use of water bottles. Figure 1.3 shows the slide being rinsed off.

5. Blot dry with *Bibulous paper* or a *paper towel*.

   (*Note:* Bibulous paper is specially prepared blotting paper with little or no extraneous paper fibers. This eliminates the sometimes annoying and confusing incidence of focusing on paper fibers under the microscope.) Figure 1.4 shows the slide being blotted dry.

6. Pour the residual stain in the staining tray into an appropriate discard container if directed to do so.

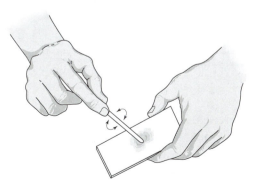

FIG. 1.1. Preparation of a smear.

FIG. 1.4. Drying a slide.

THE MICROSCOPE

A major component of this part of the course is staining and observing various types of microbes such as bacteria, fungi, and protozoans. (Viruses are too small to be seen with a conventional microscope.) Use of the microscope is an integral part of this study. These organisms, especially bacteria, are significantly smaller than any human or mammalian cells you may have seen in any Anatomy and Physiology or Biology course. Therefore, it is important that this device be used properly so that you can see the fine shapes, sizes, and structures of such small organisms. See Fig. 1.5 for a chart showing the relative sizes of microscopic structures.

The type of microscope used in most courses is a bright field, binocular, compound microscope. It is a *bright field* because it projects bright light through the image on the slide; it is *binocular* because you can use both eyes to view the object; and it is *compound* because it uses a series of lenses to achieve magnifications of up to 1000 times. The following is a basic review and operating guide for using your microscope. Gaining familiarity with the microscope components and procedures for its use will certainly enhance your use of this instrument.

Microscope Components

(See Figs. 1.6 and 1.7 which show different, labeled views of a compound microscope.)

Ocular or Eyepiece. The typical ocular has a 10X magnification; that is, it will magnify any object 10

times its size. If you are using a binocular microscope, you must make certain adjustments so that you can use both eyes to view the object on the slide. The oculars must be adjusted to account for the distance between the eyes as well as for differences in focusing ability between the right and left eye. Notice that the oculars can be moved closer and further away from each other to adjust for the distance between the eyes, and they can also be focused independently of each other to adjust for the different focusing ability of each eye. Most binocular microscopes have one fixed focus ocular and one ocular that can be adjusted. (Some binocular microscopes allow both oculars to be adjusted.)

With the adjustable lens set at the neutral ("0") position, focus the microscope at a higher magnification using only the fixed focus lens. For example, if the fixed focus lens is the right lens, look through this lens with your right eye and adjust the microscope so that the target object is in focus. Once in focus, look through the adjustable ocular and determine whether the target object is still in focus. If in focus, both eyes are able to focus at the same point in space and no further modifications are necessary. If the object looks out of focus with the adjustable lens set at "0," turn this lens clockwise or counterclockwise until the object is in focus. If you are near-sighted or far-sighted in one eye compared to the other eye, this procedure will adjust the microscope for your eyes and no eyeglasses will be needed. If you have astigmatism, however, you will still have to use eyeglasses if you want to use both oculars.

Objectives. Three objectives are typically used in this course. The 10X objective or *low-power* lens, which will give a total magnification of 100X when used with the ocular (10 × 10), is used to initially focus the object on the slide. The 40X or *high-dry* objective (total magnification of 400X) will enlarge the object 4 times more than the 10X lens. It can also be used to select an interesting field of vision, or view, before changing to the 100X or *oil immersion* objective. Finally, the oil immersion objective (1000X total magnification) is used to view all the slides prepared in this class. Most microbes are so small that even at 1000 magnifications, they will be just visible. Notice that these lenses can easily be utilized by rotating the base or *nosepiece* where they attach to the body of the microscope.

Mechanical Stage. The mechanical stage is where the slide is placed for viewing. Notice the stage clip that is used to hold the slide in place. Most microscopes have stages and stage clips that can be manipulated by turning two knobs located below the stage.

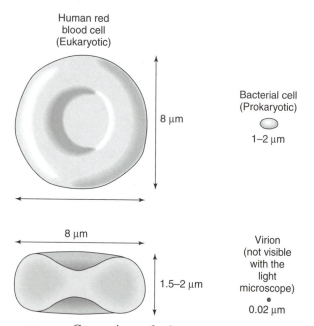

FIG. 1.5. Comparison of microscopic structures.

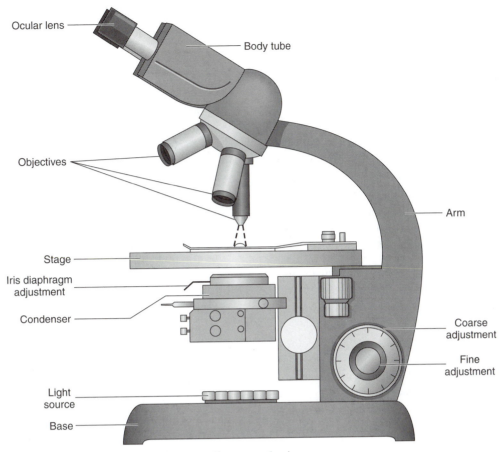

FIG. 1.6. Compound microscope.

Coarse and Fine Adjustments. These knobs are usually located on both sides of the body at or below the level of the stage. The *coarse adjustment knob* is the larger of the two and is usually located closer to the body

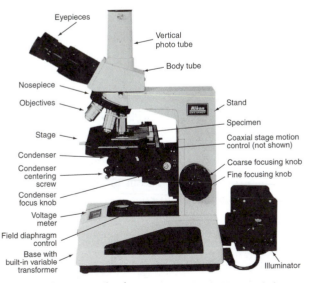

FIG. 1.7. Compound microscope. Reprinted by permission Nikon Inc., Instrument Group, Mellville, N.Y.

of the microscope. By turning it with the low power (10X) in place, you should readily see the stage move up and down unless the stage is at the highest position. **The only time the coarse adjustment knob is used is when the low power lens is in place.** If used with the higher power lenses, damage to the slide and the lens itself may occur. Most microscopes have a safety stop built into the coarse adjustment knob that prevents it from being raised too high. Once you reach this level, do not continue to turn the knob, for it may damage the microscope. Even with a safety stop, always view this adjustment of the lens and stage from the side to ensure that you do not damage any microscope components.

The *fine adjustment knob* is the smaller knob that is usually attached to the side of the coarse adjustment knob. By turning this knob, the stage will also move up and down but in much smaller increments. The movement is so minuscule that few students can see the stage move at all. One of these fine adjustment knobs may have a scale attached to it which is useful to measure the thickness of cells under the microscope. Each notch on this scale usually measures an increment of only 2 micrometers (μm), about $\frac{1}{4}$ the diameter of a human red blood cell. This knob will be the only one used to focus the microscope when using the higher magnification.

Light. Most microscopes utilize the following components to adjust the light for optimum viewing.

Illuminator Rheostat

Most microscopes make use of a rheostat to adjust the amount of electricity through the light bulb located beneath the stage. As the amount of electrical current increases, so does the illumination. Depending on the type of microscope utilized, your rheostat may be part of an on-off switch or it may operate separately. Follow your instructor's directions in operating the rheostat.

Note: To prevent premature lamp burnout, the rheostat must be turned to its lowest position before turning the microscope on and off.

Condenser

The light is focused through the slide with this lens. Its ideal position is just below its highest position beneath the stage. To adjust this lens, locate the condenser focus knob under the stage and move the lens as directed, usually to its highest point.

Iris Diaphragm

Once the rheostat and condensing lens are set, light passing through the slide can be regulated by simply adjusting the iris diaphragm located on the condenser itself. An adjustment knob or a lever is used to readily control the passage of light through the condensing lens. When using the microscope under low power (10X objective), adjust the opening of the diaphragm so that a minimum amount of light passes through the slide. As the magnification of the microscope is increased, you will have to increase the light transmitted through the slide by increasing the size of the opening of the iris diaphragm.

One of the most common problems students encounter with their microscopes is improper illumination. If you are having trouble seeing the object, the light adjustment should be one of the first things you should check.

PROCEDURE FOR USING THE MICROSCOPE

Step 1. Clean the lenses using *lens tissue* only. Start with the oculars and then the objectives. Clean the oil immersion objective last, so that any residual oil left from previous classes will not smear the oculars or other objectives.

Step 2. Set the microscope up as follows:

Have the 10X or low power objective in place.

Set the stage as high as it will go or lower the nosepiece to minimum working distance.

Set the rheostat, condensing lens, and iris diaphragm as directed.

Have the cheek cell slide, previously stained, or a prepared slide centered on the stage. (The light coming from the condensing lens acts as a spotlight to easily target the slide.)

Step 3. Look through the ocular and lower the stage. If you are looking at your own cheek smear, epithelial cells will eventually come into focus. If you are looking at a prepared slide, you will see extremely small structures. It may be advisable to focus in on the edge of the cover slip or a mark on the slide for the initial focusing. Once in focus, adjust the light and maneuver the slide so that the cells are in the center of the *field of vision.* (The circle of light you see through the ocular is the field of vision.)

Step 4. Without moving the stage, rotate the nosepiece until the high-dry (40X) objective is in place. Notice that this objective just barely misses the slide as it is rotated into place. This is why you must *use only the fine focus with the higher magnifications!* **USE OF THE COARSE ADJUSTMENT MAY DAMAGE THE SLIDE AND THE LENS!** You should notice that you can see the object, but it may be slightly out of focus. These microscopes are designed to be *parfocal;* that is, the lenses have been adjusted to focus at the same point in space. This means that if you get the object in focus under low power, it will be in focus (or nearly so) under the other magnifications. Once you use the fine adjustment to focus, you may have to adjust the light. The cells seen are now 4 times larger (400X versus 100X), and the field of vision is now 4 times smaller. Because of this smaller field of vision, a cell on the periphery of the field at 100X magnification will not be seen at 400X magnification.

Step 5. Rotate the nosepiece once again until the oil immersion (100X) is about to lock into place. Place 1 to 2 drops of immersion oil on the slide in the location where the lens will rest, and then complete the rotation and lock the objective in place. The light from the condensing lens will guide you to the exact location for placement of the oil. The objective should now be touching the oil. Under this magnification, the oil increases the *resolution* of the microscope. That is, it will give a sharp, clear image. Focus with the fine focus knob and adjust the light using the iris diaphragm.

⚠ REMEMBER NEVER USE THE COARSE
ADJUSTMENT WITH THE HIGH-DRY OR
OIL IMMERSION OBJECTIVES.

Trouble Shooting

If you are having trouble getting the object in focus with the microscope under higher magnifications, consider the following:

1. Is the light adjusted properly? If not, review the steps in adjusting the light.

2. Is the slide upside down? If it is, you will get the object in focus under low power but not with the other objectives. *Hint:* Mark a part of the slide with a marking pen or pencil for a reference point. Prepared slides will have a label and cover slip, so this will not be a problem whenever these are used.

3. Was the object in focus under low power? Remember that these objectives are parfocal. If it is out of focus in low power, it will be out of focus with the others. If you are having trouble focusing the object under low power, focus on a mark placed near the smear or on the edge of the cover slip.

4. Is the oil touching the lens? You will not get a high-resolution image with the oil immersion objective unless it is in contact with the oil.

5. Is the lens dirty? Use lens tissue to clean the lenses.

EUKARYOTIC VERSUS PROKARYOTIC CELLS

One of the many methods of classifying organisms is to divide them into two major groups based on cellular structure. *Eukaryotic cells* (*eu* = true or real, *kary* = nuclear or chromosomal material), exemplified by human cells, have characteristics that include a membrane-covered nucleus with paired chromosomes and that tend to be relatively large. *Prokaryotic cells* (*pro* = before) have no nucleus, only one chromosome, and they are small compared to eukaryotic cells. Prokaryotic cells are bacteria. Human epithelial cells tend to be 30 to 40 μm in diameter, whereas most prokaryotic cells are only 1 to 2 μm wide. The most typical prokaryotic cell found on the surface of human cheek cells is a paired circular-shaped bacterium called a diplococcus (*diplo* = paired, *coccus* = berry or round). Even with 1000X magnification, the diplococci (*cocci* = pleural of diplococcus) will just barely be visible. Figure 1.8 shows epithelial and diplococcci together. The diplococci that

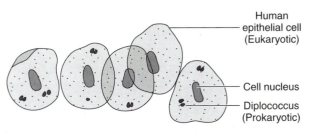

FIG. 1.8. Human epithelial cells and diplococci.

will be observed under the microscope are most likely streptococci or chains of cocci when grown in the laboratory. In the mouth, they tend to be found as pairs.

Adjust the focus and light so that the bacterial cells are as clear as possible. Notice that if you rotate the fine adjustment knob so that the focus changes only 2 μm, these cocci will no longer be in focus. If you change the amount of light going through the slide by adjusting the iris diaphragm, notice that the cocci will be much more difficult to see. This is why you must be very precise in your operation of the microscope.

While observing these diplococci, see if you can also see three slightly different shapes of these cells. All three shapes are usually found in the mouth. If you have a good smear and a good stain, and if you focus the microscope properly, you will have a good chance of observing all three on your slide. (One type will look like perfectly round pairs of cells. Another type will look like two elongated letter *D*'s back to back, while the third pair will look like two kidney beans facing each other.)

LABORATORY CLEANUP

An important part of this course is leaving your equipment and work area in proper condition for the next person to use.

Microscope Cleanup

1. Remove the slide.

2. Adjust the rheostat to dim the illuminator; then turn off the microscope lamp. As previously stated, this procedure increases the life of the bulb.

3. Clean the lenses of the microscope with *lens tissue only* in the same manner as before. Make sure the low-power (10X) objective is pointing downward. Wipe all the oil off of the 100X objective.

4. Clean the stage if necessary.

5. Wrap the power cord around the microscope.

6. Cover the microscope if a cover is available, and store the microscope in its assigned place.

Discards

Discard used and broken slides in the designated sharps container or in a container of disinfectant solution such as 10% bleach.

General Cleanup

1. Wipe your work area down with disinfectant solution.
2. Return the stains, immersion oil bottles, staining trays, prepared slides, and all other equipment and materials to their proper locations.

Speaking of Safety

The goal of every microbiology instructor is to create a safety consciousness in students that will continue to affect them in other laboratory courses, at home, and in the workplace years after the college laboratory experience is over.

A laboratory accident may seem an unlikely eventuality to many students, yet year after year hundreds of undergraduate laboratory accidents occur nationwide. Every precaution must be taken to assure a safe, enjoyable laboratory program.

⚠ REMEMBER THESE SAFETY RULES

1. No food or beverage is to be taken into a laboratory where accidental hand-to-mouth contamination or ingestion can occur. This is one of the most common ways in which dangerous microbes can enter the body.

2. Never place personal items such as pocketbooks or clothing on your laboratory table. Not only will they very likely get stained and dirty, but they also may become contaminated by microbes.

3. Wear appropriate dress: Use a lab coat, hair net, or rubber band for long hair, and put on protective gloves and eyewear when needed. Microbiological stains do one thing very well—they stain. Do not wear open shoes or apply makeup in the lab. It is better to get these stains on a lab coat than on your personal clothing. Long hair may get scorched in the Bunsen burner or pick up unwanted stains.

4. Know where the first aid kits, safety equipment, and exits are located in your laboratory. It is too late to study a floor plan when the lab is filled with fumes.

5. Follow your instructor's explicit instructions in the event that the laboratory must be evacuated. After evacuation, stay with your instructor. Never leave until you are permitted to do so.

6. Never put anything in your mouth while in the laboratory. No solutions should be pipetted by mouth. Again, this is one of the most common ways in which microbes can enter the body.

7. Never use a substance or chemical that is missing a label.

8. Always use the fume hood when instructed to do so. A chemical does not have to have an obnoxious odor to be toxic.

9. Notify the instructor in the event of a chemical spill or accident. Certain spilled chemicals will rapidly fill the lab with fumes.

10. If at all possible, do not share Bunsen burners; this can lead to singed fingers.

11. Follow laboratory housekeeping rules such as washing down your tables with disinfectant before and after use. Allow the disinfectant to air dry. For maximum effectiveness, do not towel dry the table. This will make for a clean and safe working environment.

12. Always wash your hands before leaving the laboratory room. (Are you detecting a pattern here?)

13. Replace your lab stool or chair under the table before you leave, and store your lab coat properly or place in a sealed plastic bag before you leave the lab. Make sure all materials have been returned to their proper location.

14. Never work in the laboratory unsupervised.

15. Make sure all gas jets are shut. If you smell gas, notify your instructor.

NAME _____ DATE _____ SECTION _____

QUESTIONS

1. Why are you not permitted to bring food into the laboratory?

2. Explain what can happen if you use the coarse adjustment with the oil immersion objective in place.

3. Why is the slide heated slightly once a smear has dried?

4. Name the two major different types of cells and state at least two differences between them.

5. State five probable reasons why an object may be difficult to see under a properly working microscope.

MATCHING

a. coarse adjustment   _____ holds the slide in place while on the microscope

b. fine adjustment   _____ used to initially focus the slide under low power

c. mechanical stage   _____ holds the objectives in place

d. nosepiece   _____ used to focus the microscope under higher magnifications

e. iris diaphragm   _____ used to increase and decrease the light transmitted through the slide

f. condensing lens   _____ used to focus the light through the slide

LABEL THESE COMPONENTS OF THE MICROSCOPE

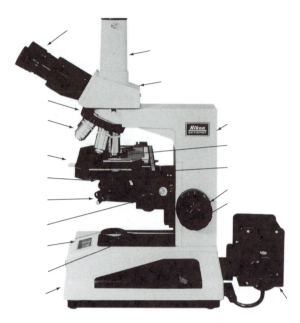

MULTIPLE CHOICE

1. The major difference between the eukaryotic and prokaryotic cell is:

    a. nuclear material
    b. amount of cytoplasm
    c. presence of a cell wall
    d. one lacks a cell membrane

2. The size of a typical bacterial cell is approximately:

    a. 2 μm   b. 8 μm   c. 30 μm   d. 50 μm

3. A microscope slide focuses properly under low power but does not do so under oil immersion. Nothing is wrong with the microscope. The most probable reason is:

    a. the light is not adjusted properly
    b. the slide was not heat fixed properly
    c. the slide was placed on the microscope upside down
    d. the coarse adjustment was not used under low power

4. Which of the following is a probable source of contamination for the microbiology student?

    a. eating in the laboratory
    b. placing personal objects on the laboratory table
    c. chewing gum in the laboratory
    d. all of these

5. The most common shape of bacteria seen in a typical cheek smear is:

    a. streptococci   b. spirilla   c. diplococci   d. single rod

6. The function of a condensing lens on a light microscope:

    a. adjusts the intensity of light going through the slide
    b. concentrates the light onto the slide
    c. magnifies the object before it reaches the ocular
    d. controls the electricity that goes through the light source

7. A parfocal microscope:

    a. uses two eyepieces
    b allows the user to view an object in three dimensions
    c. allows you to switch objectives without making major changes in focusing
    d. uses more than one lens to achieve final magnification

8. Total magnification of a light microscope is achieved:

    a. magnification power of the condenser X magnifying power of the ocular
    b. magnification power of the condenser X magnifying power of the objective lens
    c. total magnification power of all objectives added together
    d. none of these

WORKING DEFINITIONS AND TERMS

Aseptic technique Working with infectious material in such a manner as to prevent the infectious material from getting where it does not belong, or getting contaminated.

Bibulous paper Specially prepared blotting paper that contains a minimum of loose paper fibers.

Eukaryotic Cells with a true nucleus and paired chromosomes.

Parfocal Feature of most microscopes that sets the focal point of all objective lenses at the same location in space.

Prokaryotic Cells with no true nucleus (membrane covered, paired chromosomes), with only one circular chromosome.

Resolution The ability to see a small object clearly under the microscope (technically, minimal distance at which two adjacent small objects can be distinguished as separate).

Rheostat Device that controls the amount of electrical current and thus the amount of light emanating from a bulb.

EXERCISE

2

Transfer and Isolation Techniques, Microbes in the Environment

Objectives

After completing this lab, you should be able to:

1. Transfer bacteria aseptically between tubes of growth media.

2. Aseptically perform a streak plate resulting in isolated colonies.

3. Properly prepare a pour plate.

4. Determine that microbes are virtually everywhere on environmental surfaces.

TRANSFER TECHNIQUE

The type of microbe that will most often be used in this laboratory will be bacteria (singular = bacterium). Bacteria are usually grown on material known as a *growth medium* (plural = media). The chemical or nutritional composition of media is extremely varied and will be covered later in this course. Media are utilized in several different forms. The most common are *liquid broth, agar slants, agar deeps,* and *agar plates.** Agar is a chemical derived from seaweed that solidifies into a jellylike semisolid. When nutrients and other growth factors are added, many different kinds of bacteria can be grown on it. When a melted agar solution is poured in a tube, tilted, and allowed to solidify, it is called a slant. When melted agar is poured into a tube and allowed to solidify without slanting, it is called a deep. When placed in a flat dish or plate (Petri dish), it is called an agar plate.

*In the early days of microbiology, the broth was often the leftover soup from the microbiologist's most recent meal. Before agar was utilized(tradition states that its use was suggested by the wife of one such microbiologist, who used it as a food thickener), slices of boiled potato and carrots were used.

All media used in this class will be sterilized in a device called an *autoclave,* which uses steam under pressure to destroy all known infectious agents. If time permits, your laboratory instructor will demonstrate this device.

Tube-to-Tube Transfers

It is often necessary to take a sample of bacteria from one tube and place it into another. The process requires care and precision, and it is performed by following established procedures or protocols generally considered *aseptic technique.* Aseptic technique will be an integral part of every laboratory session in this course and will be utilized daily when you ultimately become a member of the medical or allied health professions.

Use of aseptic technique in the transfer of bacteria from one tube to another:

1. Prevents the microbe in the tube from *contaminating* the work area or the person performing the transfer.

2. Prevents microbes in the work area or those on the person performing the transfer from *contaminating* or mixing with the microbe being transferred.

Basic Aseptic Technique Procedures

Flaming the Inoculator. An inoculator is a thin wire attached to a handle used to aseptically transfer bacteria (the inoculum) into various types of growth media. If the end is twisted into a loop, the device is called an *inoculating loop*. If it simply comes to a point, it is called an *inoculating needle* or *stab*. You will be using the inoculating loop for most procedures in this course.

The loop is flamed before and after all procedures in order to destroy any contaminating microbes present on the loop itself. Figure 2.1 demonstrates the proper procedure for flaming the inoculator in a burner. Grasp the inoculator (loop or needle) with your dominant hand, as you would a pen or pencil. Then flame the loop and the wire to redness by holding it at approximately a 60 degree angle just outside the blue cone of the Bunsen burner flame. After flaming, the heated loop and wire should not be allowed to touch the counter top or any rubber tubing.

Note: If you are sharing a burner with another student, you must take care not to flame both inoculators at the same time. Otherwise someone may get singed.

Holding Transfer Tubes. You hold the tubes used to transfer bacteria in your nondominant hand. (If you are right-handed, use your left hand; if you are left-handed, use your left hand; and if you are ambidextrous, it's up to you.)

Hold both tubes together in your hand so that they are parallel to each other and touching. If either or both tubes have a screw cap, loosen the cap(s) to the point where they will lift right off. If either or both tubes have pop-off caps, they can usually be removed without loos-

ening them first. Make sure the tops of both tubes are level with each other.

Opening Transfer Tubes. Transfer tubes are opened by placing the caps of these tubes in the palm of your hand adjacent to your pinky, as you make a fist. (See Fig. 2.2a.) The caps can now be removed without contaminating the tubes, the caps, or your hand. *Avoid using your thumb to hold the caps.* You will be using your thumb to hold and guide the inoculator. *Never place the caps on the table.*

Alternative Method. Another method of holding and opening these tubes is to hold them so that the caps are approximately a centimeter apart from each other. Remove one cap with your pinky and the other with your ring finger. This method is useful if cotton or foam rubber plugs are used in your tubes or if your pinky is very short. (See Fig. 2.2b.)

Flaming the Tubes. With the caps removed, flame the lips of both tubes briefly to kill any airborne mi-

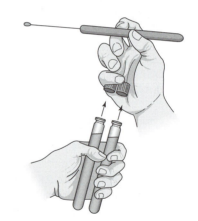

FIG. 2.2a. Holding caps with your pinky.

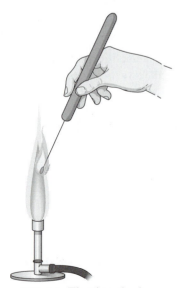

FIG. 2.1. Flaming the loop.

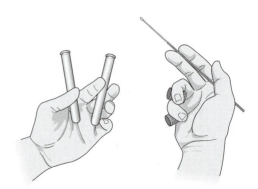

FIG. 2.2b. Alternative method of holding tubes and caps. Avoid using your thumb.

FIG. 2.3. Flaming the tubes.

crobes that may contaminate the top of the tube during transfer procedures. (See Fig. 2.3.)

Transferring the Inoculum. Once the tubes are aseptically opened and flamed, place the previously flamed loop into the tube containing the bacterial growth, withdraw some bacteria (inoculum), and place it into the sterile tube.

⚠ **CAUTION:** DO NOT ALLOW THE LOOP TO TOUCH THE LIP OF EITHER TUBE DURING THE PROCESS. MINOR MODIFICATIONS OF THIS PROCEDURE CAN BE MADE DEPENDING ON WHETHER YOU USE BROTH TUBES OR AGAR SLANT TUBES. (SEE FIG. 2.4.)

Finishing the Transfer. Remove the loop from the newly inoculated tube. Again flame the lips of both tubes briefly and re-cap them. Flame the loop to redness before you put it down.

Four slightly different procedures are used in transferring the *inoculum,* or sample of microbe, from one tube to another. They are: broth to broth, broth to slant, slant to broth, and finally, slant to slant. Once mastered, they can be modified for other nonlaboratory procedures such as taking cultures from patients. In performing these four different transfers, you will note that almost

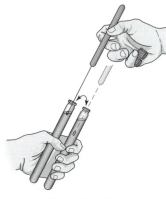

FIG. 2.4. Transfer of inoculum.

all the steps are exactly the same for each procedure. In other words, if you can perform one of these procedures, you should have little difficulty performing the others.

⚠ **SAFETY RULE:** EACH PERSON WILL USE HIS OR HER OWN BUNSEN BURNER WHENEVER POSSIBLE. AVOID PLACING IT IN FRONT OF YOUR TEST-TUBE RACK. IF YOU HAVE LONG, LOOSE HAIR, WEAR A HAIR NET OR USE A RUBBER BAND TO HOLD YOUR HAIR AWAY FROM THE OPEN FLAME.

Practice the four procedures using sterile tubes of broth and slants. Do not use any of the provided cultures of bacteria until instructed to do so.

Microbes Per Table/Workstation:

broth culture of *Serratia marcescens;* slant culture of *Sarcina flava* or *Micrococcus luteus*

Materials List Per Student/Workstation

Two tubes of sterile broth

Two tubes of sterile agar slants

Inoculating loop

Test-tube rack

Marking pen or pencil

Striker (for lighting Bunsen burner)

Broth-to-Broth Transfer

Note: Always carry or store broth tubes upright. If tilted too much, the broth will reach the cap and spill, or (possibly) become contaminated.

1. Hold both tubes in your left hand (if you are right-handed). If you are using screw-capped tubes, loosen both screw caps to the point where they will lift right off.

2. Grasp the inoculator (inoculating loop) with your right hand as you would a pen or pencil. Flame the loop to redness by holding it at an approximately 60 degree angle, just outside the blue cone of the Bunsen burner. (See Fig. 2.1.)

3. Open both tubes by placing the caps in the palm of your hand adjacent to your pinky, and making a fist; alternatively, hold the caps slightly apart and use your pinky and ring finger to remove the caps. (See Figs 2.2a and 2.2b.) *Avoid using your thumb and never place caps on the table!*

4. Flame the lips of both tubes briefly to kill any airborne microbes that may contaminate the top of the tube during the transfer. (See Fig. 2.3.)

5. Carefully place the loop into the tube with bacterial growth, mix briefly, and carefully remove the loop without touching the rim of the tube.

Note: Do not allow the loop to touch the lip of the tube at any time. Since it is rather springy, it can spray the inoculum over you and your work area.

6. Place the loop in the sterile broth and mix gently for a few seconds.

7. While withdrawing the loop, tap it gently on the inside of the tube well below the lip. This removes any residual broth from the loop.

8. After the loop is withdrawn, flame the tubes again and replace the caps.

9. Flame the loop to redness and place it on the test-tube rack. Take care not to contaminate the loop by placing it on the tabletop. Even though the loop was just sterilized by placing it in the Bunsen burner, it must be resterilized before it is used again.

10. If you are using screw-capped tubes, tighten the screw caps until they are snug and then loosen a quarter turn. This allows oxygen to enter the tubes. If you are using pop-off caps or cotton plugs, simply push the caps over the end of the tube or the cotton plugs into the top of the tubes.

Broth-to-Slant Transfer

Note: Always carry or store broth tubes upright. If tilted too much, the broth will reach the cap and (possibly) become contaminated.

1–4. Repeat these steps as you did for the broth-to-broth transfer.

5. Carefully place the loop in the broth tube, mix briefly, and carefully remove the loop without touching the rim of the tube.

⚠ **CAUTION:** DO NOT ALLOW THE LOOP TO TOUCH THE LIP OF THE TUBE AT ANY TIME.

6. Place the loop in the lower part of the sterile agar slant, touch it to the agar, and draw it gently, running up the surface of the slant once while still touching the agar.

7. After withdrawing the loop, flame the tubes again and replace the caps.

8. Flame the loop to redness and place it on the test-tube rack. Take care not to contaminate the loop by placing it on the tabletop.

9. If you are using screw-capped tubes, tighten the screw caps until they are snug and then loosen a quarter turn. This allows oxygen to enter the tubes. If you are using pop-off caps or cotton plugs, simply push the caps over the end of the tube or the cotton plugs into the top of the tubes.

Slant-to-Broth Transfer

Note: Always carry or store broth tubes upright. If tilted too much, the broth will reach the cap and (possibly) become contaminated.

1–4. Repeat these steps as you did for the broth-to-broth transfer.

5. Carefully place the loop in the slant, cool it by touching a sterile part of the slant, then pick up some bacterial growth, and carefully remove the loop without touching the rim of the tube.

⚠ **CAUTION:** DO NOT ALLOW THE LOOP TO TOUCH THE LIP OF THE TUBE AT ANY TIME.

6–10. Repeat these steps as you did for the broth-to-broth transfer.

Slant-to-Slant Transfer

1–4. Repeat these steps as you did for the broth-to-broth transfer.

5. Carefully place the loop in the slant, cool it by touching a sterile part of the slant, then pick up some bacterial growth, and carefully remove the loop without touching the rim of the tube.

⚠ **CAUTION:** DO NOT ALLOW THE LOOP TO TOUCH THE LIP OF THE TUBE AT ANY TIME.

6–9. Repeat these steps as you did for the broth-to-slant transfer.

When directed by your instructor, perform the same procedure using living bacteria as the source of inoculum. Place the practice tubes aside in the test-tube rack and get two more sterile broth tubes and two more sterile slants. Make sure you gently mix the inoculum in the broth tubes before the transfer. This is accomplished by tapping the bottom of the tube with your finger, shaking gently, rolling the tube in the palm of your hand, or even using a vortex mixer if available. Label the tubes with a pencil or permanent marker so that you know which microbe is in which tube and, of course, which of the tubes are yours.

⚠ REMEMBER

1. Gently mix the inoculum in the broth tubes before transfer.

2. If screw caps are used, loosen the screw cap a quarter turn before placing the tubes in the container for incubation.

3. Label the tubes so that you know which microbe is in each tube and, of course, which of the tubes are yours.

ISOLATION TECHNIQUES: STREAK PLATE AND POUR PLATE (OPTIONAL)

The microbiologist must be certain that any microbe used or studied in the laboratory has not been contaminated with others from the environment or from the microbiologist himself. Therefore, the microbiologist must constantly make sure that he or she is working with a *pure culture;* that is, every microbe in that tube or plate must be exactly the same. In order to make sure you have a pure culture, or to get a pure culture from a mixture of different microbes, certain techniques have been developed to separate individual cells from large numbers of microbes, and to allow these cells to grow into pure cultures. The German physician Robert Koch was the first to develop one of these techniques, and the following procedure is similar to the methods he used. The purpose of these techniques is to get *isolated colonies* from a large number of different microbes. (A colony is growth resulting from a single microbe placed on an agar surface, well separated from other microbes. Within each colony, the cells are genetically identical.)

Streak Plate

The streak plate is the most popular and easiest method of getting isolated colonies from large numbers of different bacteria. The procedure for streaking a plate for isolated colonies involves gently drawing a loopful of inoculum numerous times across the surface of an agar plate, thus placing streak marks on the surface of the agar. Initially, hundreds, even thousands, of individual cells are placed on the agar plate. The streaking is done in such a way as to "thin out" the microbes so that, eventually, only one bacterial cell at a time is placed on the plate, well separated from the others. When allowed to reproduce, huge numbers of these bacterial cells grow together into visible isolated colonies.

Cultures/Table or Work Area:

broth culture of *Serratia marcescens;* agar slant culture of *Sarcina flava* or *Micrococcus luteus;* agar plate culture of *Bacillus subtilis*

Materials List Per Student/Workstation

Two or three practice plates

Three nutrient agar or trypticase soy (T-Soy) plates

Inoculating loop

PROCEDURE

Practice the streak plate technique by first using a sterile inoculating loop and two or three practice plates.

1. Place the agar plate *upside down,* that is, agar side up with lid side down, on the table in front of you.

2. Flame the loop to redness, allow it to cool, simulate taking a sample of inoculum from a tube or another agar plate (may be omitted for practice), and pick up the bottom of the plate in the palm of your hand.*

3. Hold the plate so that light from above shines off the surface of the agar, and gently place the loop on the edge of the plate. The loop should be at a 30 to 45 degree angle with the plate.

4. Gently draw the loop across the surface of the agar in a zig-zag pattern in such a way as to avoid overlapping the previous streak. With practice, you will be able to accomplish this without tearing into the agar. By holding the plate so that light is shining off the agar, you can determine the exact position of the loop.

---

*An alternative to this method is to keep the plate right side up, lift the cover slightly, and streak with the cover only slightly ajar. This prevents dust (and microbes) from the air from landing on the plate. Most modern microbiology labs have efficient filtration systems, so this practice is rarely used today. However, if your laboratory area happens to be particularly dusty or the windows are open, this method may be an option.

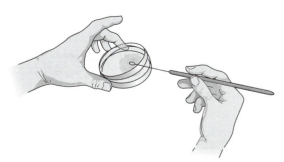

Alternate method of performing a streak plate.

5. Cover approximately one-fourth to one-third of the plate with between 10 and 20 streak marks. (See Fig. 2.5.) Do not overlap previous streaks.

6. *Flame the loop*. This removes any bacteria from the first section of streaking. Allow the loop to cool by touching it to a sterile section of the agar plate or by waving it in the air.

7. Draw the loop diagonally across the first group of streak marks *once,* thus picking up a small number of microbes from the first section of the plate.

8. Cover a second one-fourth to one-third of the agar plate using the same technique as in steps 4 and 5 above. (See Fig. 2.6.)

9. Flame the loop again and repeat the process until all three or four sections of the plate are covered with streak marks. (See Figs. 2.7 and 2.8.) (Professional microbiologists are often able to achieve isolated colonies without flaming between each section. However, until you reach such proficiency, you should flame the loop at least once during this procedure.)

10. Clinically, growth is often reported by the presence of growth using four quadrants (i.e., "quadrant 1 growth" or "quadrant 4 growth").

11. Replace the bottom of the plate back in the cover and flame the loop.

Note: if you cut into the agar during practice, or even during a real inoculation, simply continue streaking using less pressure, or change the angle of the loop to the agar. It is poor technique to dig into the agar, but with practice, you will avoid this error.

⚠ REMEMBER

1. Keep the plate inverted (upside down) before and after inoculation and during incubation. This prevents any moisture that may have accumulated on the inside of the cover from dropping onto the surface of the agar and prohibiting proper isolation.

2. Flame the loop at least once during the streaking procedure.

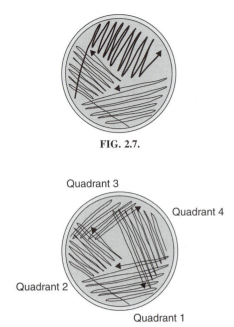

FIG. 2.7.

Quadrant 3

Quadrant 4

Quadrant 2

Quadrant 1

FIG. 2.8. Plate inoculated in three or four quadrants. In clinical specimens, if a sample is taken directly from patients, this gives the microbiologist a general idea as to the concentration of the inoculum. This procedure will not work for laboratory specimens as these specimens are already highly concentrated after being allowed to grow under ideal conditions.

3. Do many streaks per section of plate, and do not overlap streaked areas.

4. Streak at a 30 to 45 degree angle to the surface of the plate. A higher angle will allow the loop to cut into the agar, whereas a lower angle will tend to smear the streaks, making it more difficult to get isolated colonies.

Prepare the Following Streak Plates Using the Cultures Provided:

From *Serratia marcescens* broth

From a *Micrococcus luteus* or *Sarcina flava* slant

From a *Bacillus subtilis* plate

In preparing these streak plates, aseptically retrieve the inoculum from the tubes and return the tubes to the test-tube rack before streaking the plates. When getting the culture material from the agar plate, make sure both agar plates are upside down on the table. Take the sample from one plate and recover the sample plate before streaking the sterile plate. Label the three plates on the bottom (not on the lids, to prevent mixing up the information) so that you know which bacteria are on which plate and which plates are yours. Ideally, the label should include the type of medium, the date, the name of the organism inoculated, and your initials. Remem-

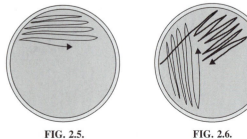

FIG. 2.5.   **FIG. 2.6.**

ber, the label should be placed on the bottom of the plate, not on the lid.

Pour Plate

Another method sometimes used to isolate bacteria is the pour plate. The pour plate also has the advantage of allowing you to know how many microbial cells were placed in the plate as you can count the colonies after growth occurs. A large test tube of melted, sterile agar growth medium (called an *agar deep*) is used for this procedure. Agar has the unique property of melting at 100° C but not solidifying until its temperature drops to 45° C. Melted agar deeps are kept ready to be used for pour plates by placing them in a waterbath maintained at a temperature slightly above 45° C. Thus, once they are removed from the waterbath, they will start to solidify within a few minutes. If these tubes are kept and used at a temperature much higher than this, the microbes placed in the tube will quickly be killed, which will nullify the purpose of this part of the laboratory exercise. You must therefore be prepared to perform the following procedure immediately after the tube is removed from the waterbath.

Cultures/Table or Work Area

Broth culture of *Serratia marcescens* or a mixed broth culture of *Serratia marcescens* and *Micrococcus luteus*

Materials List Per Student/Workstation

Melted agar deep

Sterile plastic Petri dish

Inoculating loop

PROCEDURE

1. Place a sterile, empty Petri dish on the table, lid side up.
2. Make sure the broth culture is readily available.
3. Get a melted agar deep from the waterbath and transfer a single loopful of the inoculum to the melted agar deep using the broth-to-broth transfer technique previously practiced. You must do this quickly because the melted agar will soon begin to solidify.
4. You may now:
   a. Mix the melted agar/bacteria mix by gently shaking (see Fig. 2.9) or tapping the tube or rolling the tube in the palm of your hand, then pour the mixture into the Petri dish (see Fig. 2.10), or
   b. Pour the melted agar/bacteria mix immediately into the Petri dish without mixing—you will see why later.

5. If the melted agar does not completely cover the bottom of the dish, gently swirl the agar first by replacing the cover and then by rotating the dish. (See Fig. 2.11.)
6. Allow the agar to solidify for 5 minutes, turn the plate upside down, and label. Incubate the plate in the inverted (agar side up) position.

When performed properly, a pour plate will accomplish the following:

1. Isolate small numbers of bacteria into colonies.
2. Provide a number of how many microbes were placed in the melted agar. (This is often called a plate count.)

Unless a highly diluted broth solution was used for this inoculation, you will not likely get isolated colonies

FIG. 2.9.

FIG. 2.10.

FIG. 2.11.

or an accurate count of the bacteria inoculated in the melted agar deep. This has nothing to do with your technique; rather, it is because you will be placing too many bacterial cells in the tube in the first place. You should see huge numbers of colonies on the Petri dish at the next lab session. It should be readily apparent that only one loopful of inoculum will be necessary to ensure growth.

MICROBES IN THE ENVIRONMENT

The aseptic technique involves procedures that are performed to keep extraneous (contaminating) microbes out of a work area. Such work areas include the sterile field in an operating room, an injection site, or the test tubes and agar plates used for bacterial growth. These extraneous microbes can be found nearly everywhere in the environment surrounding the work areas. Usually, the major source for such microbes is the person who actually performs the aseptic procedure. This part of the laboratory will demonstrate the omnipresence or ubiquitousness of microbes in the environment. In other words, unless some procedure was performed to eliminate or reduce microbes in an area, you can safely assume that microbes will be present.

Materials/Student

Blood Agar Plate (if available)

T-Soy Agar Plate

Sterile swabs

Sterile water blanks

Human Environment: Procedure

The human body is a major source of microbes, which can often contaminate sterile materials, work surfaces, and even patients. You can determine that the human body is a source of bacterial contamination by taking a sample from your body and placing that sample on the surface of a blood agar plate (or T-Soy plate if the blood agar is not available).

The sample can be obtained in one of three ways:

1. Press the surface of the agar directly on an external part of the anatomy (e.g., forehead, hair, hand, or elbow).
2. If the sample is to be taken from a moist area that cannot be touched directly (throat, gums), use a sterile cotton swab to obtain the sample and spread the swab across the blood agar plate. (In a later labora-

tory, you will be shown how to get isolated colonies from such an inoculation technique.)
3. If the sample is to be taken from a dry area such as between the fingers or the ear, moisten the cotton swab with sterile water before obtaining the sample.

Classroom/School Environment: Procedure

Perform the same procedure described above using a T-Soy agar plate. If the sample you choose to test can be pressed on the agar plate directly (e.g., lab coat), do so. If it is moist but will not fit on the plate directly (e.g., faucet, sink, moisture on or in a refrigerator), use a sterile cotton swab. If the chosen sample is dry, first moisten a sterile swab with sterile water.

Inventory

After completing this exercise, you will have the following tubes and plates ready for incubation:

Two broth tubes, properly labeled (Remember, loosen screw caps.)
Two agar slants, properly labeled
Three streak plates, one from broth, one from a slant, and one from another agar plate
One pour plate
Two environmental sample plates

Note: Make sure the plates are inverted upside down. If agar plates are incubated lid side up, moisture from the lid may drop on the surface of the plate during incubation. This will allow bacteria to spread, thus effecting colony isolation.

Results

Observe the transfer tubes you have inoculated after they have been allowed to grow. Although it is proper technique to gently shake the broth cultures as part of the procedure, *do not do so at this time*. You should notice obvious growth in the slants, and you will probably see a precipitate in the broth.

Now inspect the streak plates. Somewhere on the surface of the plate, you should see well-isolated colonies. It doesn't matter where these colonies are located, just so they are separated from each other. If you do not see isolated colonies, see the instructor for suggestions on how to improve your technique.

Note: Bacillus subtilis produces large colonies. If you observe well-isolated colonies in this streak plate, you have mastered the technique.

Set aside several minutes during future laboratory sessions and practice your transfer and streak plate techniques in order to improve and maintain these skills.

Observe the pour plate you prepared. Use a magnifying glass or stereo microscope if available. The huge numbers of small specks seen in and on the agar are individual colonies of the bacteria you inoculated from a single loopful of broth. There will probably be too many colonies to either isolate or count unless your instructor diluted the broth culture beforehand. (This is the usual function of a pour plate.) You should now realize that even a small loopful of broth is quite adequate to ensure growth when incubation procedures are properly followed. If you poured the agar/bacteria mix directly into the agar plate without mixing first, you will probably see a very artistic mosaic arrangement of the colonial growth. Other than the aesthetic beauty of viewing such growth, there is no medical or scientific need for following this technique.

The two environmental sample plates will demonstrate that there are microbes on (or in) you and on virtually all other surfaces with which you have come into contact. By observing these plates, you can determine whether the sample was taken from an area of high or low microbial concentration. Notice that almost all the plates probably have some growth on them. This means that you must always follow aseptic technique procedures. If you happened to choose two sites with relatively low microbial concentrations, look at the results from other environmental plates in the classroom.

LABORATORY CLEANUP

Incubation

Most bacterial cultures grow best at a temperature of 35° C, just 2° C below that of the human body. A device known as an *incubator* is used, which accurately maintains this or any other temperature that it is set for. If tubes are to be placed in the incubator, make sure the screw caps are slightly loosened and that all tubes are properly labeled. Agar plates must also be labeled properly and placed in the incubator or incubation tray *lid side down*. Inverting the plates prevents moisture that may be present on the inside cover from splashing down on the developing colonies. (This moisture forms a temporary broth solution, which allows the previously isolated bacteria to spread all over the surface of the plate.) After an incubation period of 18 to 24 hours, these cultures are inspected or placed under refrigeration for later observation. Refrigeration impedes further growth of the bacteria, so you will have "fresh" cultures to work with during the next laboratory session.

Discards

Discard all tubes and plates that are not placed in the incubation tray or incubator. Follow the lab instructor's direction for tubes and agar plates. The discards will include your cultures from the previous week, as well as any tubes distributed to you or your table during the current lab.

General Cleanup

Return loops, test-tube racks, and all other equipment and materials to their proper location. Clean the tabletop with disinfectant, and place the stools or chairs under the table. Properly store your lab coat in the designated area, or place it in a plastic bag before leaving the lab.

NAME _____ DATE _____ SECTION _____

QUESTIONS

1. Why would you be instructed not to share Bunsen burners?

2. What is the reason for flaming the tubes before and after each transfer?

3. Why would you avoid using your thumb to hold the caps during a transfer?

4. Explain why you should avoid allowing the loop filled with inoculum from touching the lip of either the source tube or the tube to be inoculated.

5. When getting inoculum from a slant, why is it necessary to touch a sterile part of the agar with the loop before touching the bacterial growth?

6. Why should the loop be flamed at least once during the streak plate procedure?

7. Why are agar plates kept inverted whenever possible?

8. What is one advantage of a pour plate over a streak plate?

9. When you inspect the environmental plates, you will probably notice that there is more growth from a sample taken from a moist area than from a dry area. Why?

10. Why didn't you get isolated colonies from the pour plate if the broth culture was not diluted first?

11. What are some of the reasons for not getting isolated colonies from a streak plate?

MATCHING

a. colony

b. pure culture

c. agar

d. broth

e. pour plate

f. inoculation

g. incubator

h. autoclave

_____ device that uses pressurized steam at 121° C

_____ a general term for most liquid growth media

_____ method of separating, isolating, and counting bacteria

_____ microbial growth in a container where all the cells are of the same type

_____ device that maintains a constant temperature

_____ a separated, visible clump of bacteria growing on an agar plate

_____ solidifying agent used for growth media derived from seaweed

MULTIPLE CHOICE

1. The most common method of achieving isolated colonies is the:
   a. broth dilution   b. agar slant   c. streak plate   d. agar deep

2. Which of the following can lead to contamination?
   a. forgetting to flame the loop between inoculations
   b. allowing the broth to reach the top of the tube
   c. allowing moisture from the cover of an agar plate to leak onto the agar
   d. all of these

3. A method for estimating the number of bacteria in a sample of inoculum is:
   a. pour plate   b. streak plate   c. broth culture   d. slant culture

4. Agar melts at:
   a. 10° C   b. 40° C   c. 60° C   d. 100° C

5. A procedure that allows a laboratory worker to properly handle microbes safely is:
   a. sterilization   b. aseptic technique   c. disinfection   d. antisepsis

6. A student prepares a streak plate of a bacterial culture, and there is very poor isolation of colonies. The reason for this failure could be:
   a. student failed to flame the loop between sections
   b. student didn't do enough streaks in each section
   c. student overlapped areas previously streaked
   d. all of these are possible

7. Aseptic technique involves:
   a. preventing of contamination of student work areas
   b. keeping bacteria found on the student from getting into cultures
   c. reducing the chances of bacteria used in student cultures from contaminating the student
   d. all of these

8. Flaming an inoculation loop before and after tube-to-tube transfers:
   a. cleans the loop
   b. prevents contamination of cultures and work areas
   c. removes toxic chemicals that accumulate on the loop
   d. allows the metal to adhere more effectively to the inoculum

9. Moisture on the inside cover of an agar plate can cause contamination. This is prevented by:
   a. keeping the plate inverted whenever possible
   b. wiping the excess moisture away using a sterilized towel
   c. placing the plate in an incubator before it is used
   d. flaming the inside cover to evaporate the excess moisture

WORKING DEFINITIONS AND TERMS

Agar Solidifying agent used for growth media derived from seaweed.

Agar deep Growth medium in a test tube allowed to solidify as the tube sits in a test-tube rack

Agar plate Solidified growth medium in a lid-covered, flat dish.

Agar slant Solidified growth medium in a test tube, allowed to solidify at an angle, thus presenting a large surface area to allow for growth.

Autoclave Device that uses pressurized (15 lbs/in^2) steam at 121° C to kill all known infectious agents in 15 minutes.

Broth General term for liquid growth medium.

Colony A visible clump of bacteria growing on an agar plate, separated from other areas of growth.

Contamination Presence or possible presence of microbes in an area where they do not belong.

Incubation (In the laboratory) allowing a microbe to grow at a constant (usually optimal) temperature.

Inoculating loop Device used to aseptically transfer and streak microbes in the laboratory.

Inoculation (In the laboratory) the process of introducing microbes into a culture medium.

Inverted The position of a Petri dish or agar plate whereby the lid side is facing downward.

Petri dish A flat dish, with a lid. The base is filled with solidified growth medium used to isolate and grow bacteria.

Pour plate Method of separating, isolating, and counting bacteria by placing a small sample of the microbe in a melted agar and pouring into a Petri dish.

Pure culture Microbial growth in a container where all the cells are of the same type (genus and species).

Slant See agar slant.

PART II

MICROBIAL MORPHOLOGY, DIFFERENTIAL STAINS

Carl Linnaeus, (1707–1778), an eighteenth-century Swedish botanist, specialized in classifying all known types of living organisms. When he decided to classify microbes, he observed some through his microscope, quickly gave up, classified everything he saw as Genus *Chaos,* and moved on to other projects.

Today, we are somewhat more sophisticated in classifying and identifying the myriad numbers of microbes in our world. A useful starting point is determining whether the microbes are bacteria, algae, fungi, protozoa, or microscopic multicellular parasites. (As noted earlier, viruses are too small to be seen with the conventional microscope available to most microbiology students.) Bacterial shape or morphology, as well as the type of protective coverings such as cell walls and capsules, are important criteria in classifying and identifying bacteria. Similar techniques are also used to categorize fungi and protozoans.

Various staining techniques have been developed to aid in these identifying procedures. When properly employed, one can easily categorize these organisms into the basic groups used in microbiology.

EXERCISE 3

Cultural and Cellular Morphology

Objectives

After completing this lab, you should be able to:

1. Properly prepare bacterial smears from broth and from agar for staining.

2. Distinguish between different bacterial morphologies or growth characteristics on agar plates, slants, and in broth.

3. Recognize the following morphological shapes and arrangements under the microscope: diplococci, streptococci, sarcinae, staphylococci, single rods, and spirilla.

4. Recognize budding in yeast cells.

CULTURAL CHARACTERISTICS OF BACTERIA

Just as different plants and animals have various *morphologies* or shapes, so do bacteria, both macroscopically and microscopically. Although you may not be required to memorize the terminology associated with the unmagnified forms of microbial growth seen in tubes or plates, it will be important for you to distinguish between different microbes by their forms of growth on various types of media.

Materials List Per Table/Workstation

Broth or agar cultures of *Serratia marcescens, Sarcina flava* or *Micrococcus luteus, Bacillus subtilis, Staphylococcus aureus, Enterococcus faecalis, Escherichia coli, Moraxella catarrhalis, Saccharomyces cerevisiae, Mycobacterium phlei, Pseudomonas aeruginosa, Proteus mirabilis* or *vulgaris*

Magnifying glass or stereo microscope if available

Unknown—from environmental plates (Exercise 2)

Crystal violet or methylene blue stain

Marking pen or pencil

Bibulous paper

Prepared slides of appropriate bacillus, coccus, and spirillum species

PROCEDURE: PLATES

During this laboratory, observe the different types of growth patterns of both isolated colonies (agar plates) and nonisolated bacteria (slants and broths). Use the three streak plates prepared from Exercise 2, the two environmental plates, as well as the plates assigned to each table. Using Fig. 3.1 as a guide, you should soon determine that most of these microbes are easily distinguished from the others. In some cases, you even get a clue as to how each bacterium got its name (e.g., *Bacillus* = rod, *luteus* = yellow).

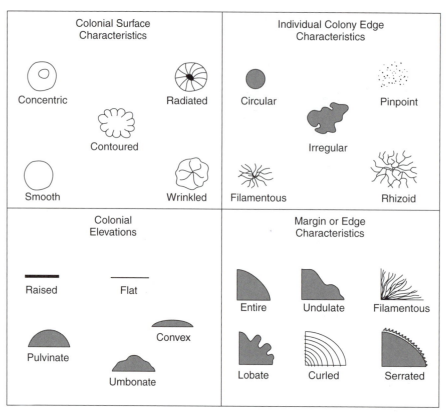

FIG. 3.1. Colonial characteristics.

PROCEDURE: AGAR SLANTS

Note that many of the bacteria provided for you and the ones you inoculated in Exercise 2 can be differentiated from each other on the basis of the type of growth on the agar slant. Observe Fig. 3.2, which shows some of the different forms of growth seen on agar slants, and compare this with the types of growth seen in the tubes at your table or workstation.

PROCEDURE: BROTH

Before observing the growth characteristics in the broth tubes at your table, review the types of growth seen in Fig. 3.3. Unless you have a fresh, 24-hour culture right from the incubator, you will not see many tubes with turbidity. This is because the bacteria tend to settle out or to precipitate when they are refrigerated for several days. After you note that either a *pellicle* or *ring for-*

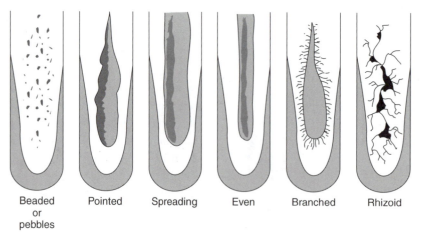

FIG. 3.2. Agar slant characteristics.

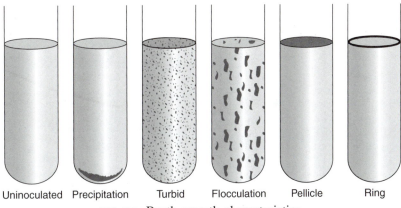

Uninoculated   Precipitation   Turbid   Flocculation   Pellicle   Ring

FIG. 3.3. Broth growth characteristics.

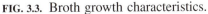

mation is possible on the surface of the broth, pick up the tubes and look for these phenomena, as well as for precipitation. Once this step is completed, mix the bacteria by gently shaking, tapping, or rolling the tube in the palm of your hand. **DO NOT SHAKE THE TUBE IN SUCH A WAY AS TO CONTAMINATE THE CAP WITH THE BROTH.** Once mixed, you may observe turbidity, flocculence, or even a ropelike appearance.

MICROBIAL CELLULAR MORPHOLOGY

Another method of distinguishing between microbes is to view them under the microscope. There are three main groups of bacteria based on individual cellular morphology: coccus (plural = cocci), bacillus (plural = bacilli), and spirillum (plural = spirilla). In addition to these three general cell shapes, many of them can be further distinguished by their cellular arrangements. Figure 3.4 shows the following group morphologies:

Rod or bacillus

    coccobacillus

    vibrio or comma

    single rod

    streptobacillus (*strepto* = chain)

    cording

    snapping, palisades or picket fence, "Chinese letters"

Coccus (berry shaped)

    diplococcus (*diplo* = pair)

    tetrad—packet of 4

    streptococcus—chain of at least 4

    sarcinae (*sarcinae* = packet of 8)

    staphylococcus (*staphy* = bunch of grapes)

Spiral

    spirillum (wavy)

    spirochete (coil or corkscrew)

SMEAR PREPARATION

Divide a slide into four sections each, using a marking pen or pencil. The section on the left will serve as a handle during the staining process and can be used to place a label afterward. The other three sections will be used to place the smears of each bacterial specimen. (See Fig. 3.5.) Once labeled, each slide can be used as a reference for future observation.

Note: Wax pencil markings and "permanent" markers will not be very permanent with many staining procedures. These markings should remain after performing this simple stain. With future stains, if a permanent label is required, it will have to be added after the staining process is completed.

BROTH PREPARATION

1. To prepare a smear from *broth,* you gently mix the tube of broth and *aseptically remove* a loopful of inoculum from the tube (loosen cap, flame loop, remove cap, flame top of tube, get inoculum, flame top of tube, replace and tighten cap, loosen a quarter turn, and return tube to test-tube rack). See Exercise 2 for a review if necessary.

2. Touch the loop to one of the three sections on the slide and spread the broth over an area at least the

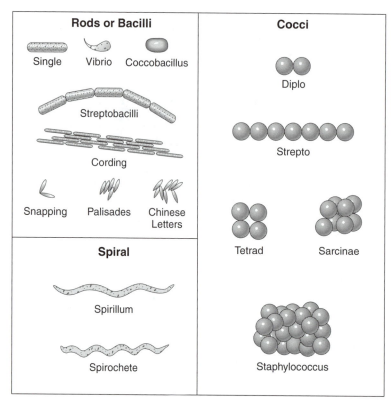

FIG. 3.4. The most common bacterial shapes.

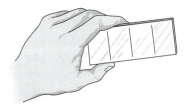

FIG. 3.5. Slide divided into four sections.

diameter of a dime. You may be directed to draw a circle under the slide where the smear is prepared to help you "target" the exact area. (See Fig. 3.6.)

3. Flame the loop and allow the smear to air dry before heat fixing. See the Simple Stain Technique in Exercise 1 for a review if necessary.

AGAR SLANT AND PLATE PREPARATION

1. To prepare a smear from agar, you must first place a small drop of water on the slide. This is easily accomplished by placing the loop in water and then touching it to the slide. The water doesn't have to be sterile, for any microbes that may be present will be lost among the thousands placed on the slide

from the slant or plate. Avoid using a large drop of water because it will take a much longer time to dry.

2a. If you use a slant, follow the same procedure in getting the inoculum as you did for the broth smear. Attempt to get a very small sample of bacteria on the loop to mix with the water. You want the smear to be "thin" or appear slightly cloudy, not as if you painted it on.

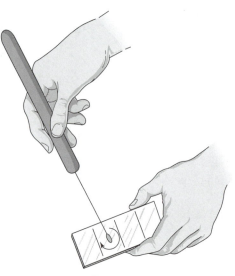

FIG. 3.6. Smear preparation.

2b. If you use an agar plate for the source of inoculum, follow the procedure for a plate-to-plate transfer up to the point where the inoculum is on the loop and the source plate for the bacteria is closed. Once again, you need a thin smear.

⚠ **CAUTION:** REMEMBER TO FLAME THE LOOP BE-TWEEN EACH INOCULATION, OR YOUR SMEARS WILL BECOME CONTAMINATED.

3. Touch the small drop of water on the slide with the sample on the loop and rotate a few times so that it is evenly distributed throughout the smear. The smear should be at least the size of a dime. Avoid excessive mixing of the inoculum, for this may break up certain groupings of bacterial cells such as streptococci.

4. Allow it to air dry before heat fixing.

SIMPLE STAIN PROCEDURE

Prepare smears of the microbes assigned by your instructor. See the chart below for recommended samples.

1. Place the air-dried, heat-fixed slide on the staining tray and cover it with crystal violet or methylene blue.

2. Stain for 5 seconds if crystal violet is used or for 1 minute if methlyene blue is used. (Review the procedure from Exercise 1 if necessary.)

3. Rinse with water and blot with Bibulous paper or a paper towel.

4. If directed, pour the residual stain in the staining tray into an appropriate discard container.

Place the prepared slides under the microscope and observe each smear under oil immersion. Maneuver the slide to the edge of the smear so that the field of vision shows well-separated groups of cells before you determine cell and group morphology. (See Fig 3.4.)

Results from the Simple Stain Procedure

FROM SIMPLE STAIN PROCEDURE

Microbe	Morphology Name	Drawing
Micrococcus luteus or Sarcina flava		
Bacillus subtilis		
Serratia marcescens		

Microbe	Morphology Name	Drawing
Staphylococcus aureus		
Enterococcus faecalis		
Moraxella catarrhalis		
Saccharomyces cerevesiae		
Escherichia coli		
Unknown (from environmental plates)		

Note: You cannot determine group morphology unless you observe the microbes on a section of the smear that shows the cells well separated from the others. Otherwise, all cocci look like staph and all rods look like snapping or Chinese letters. Even if your smears are somewhat thick, there is usually someplace where there is enough separation between cells to accurately determine group morphology.

Results from Prepared Slide

FROM PREPARED SLIDE

Microbe	Morphology Name	Drawing
Bacillus		
Coccus		
Spirillum		

Inventory

After completing this exercise, you will have done the following:

Several practice transfer tubes and streak plates ready for incubation (continuation of practicing skills from Exercise 2)

Three slides of bacterial and yeast smears prepared with a simple stain

LABORATORY CLEANUP

1. Place your labeled slides in the assigned slide boxes or discard as directed. If instructed to do so, save your slides. Do not attempt to wipe off the oil, for this will remove the smear.

2. Clean and put away the microscope as described in Exercise 1.

3. Put away all stains, staining trays, and loops.

4. Clean the tabletop with the disinfectant solution and place your stool or chair under the table.

5. Incubate the tubes and plates as described in Exercise

2. Remember to loosen the caps on any screw-capped tubes.

6. Discard all tubes and plates not to be incubated as described in Exercise 2.

7. Clean up as described in Exercise 2.

NAME _____ DATE _____ SECTION _____

QUESTIONS

1. Why must a smear be made thin to show proper bacterial microscopic morphology?

2. What is the procedural difference between using broth and agar (slant or plate) as a source of bacteria for preparing a smear?

MATCHING

a. pointed, spreading,
   pebbled, even

b. pinpoint, circular,
   filamentous, irregular

c. flocculent, ring formation, pellicle, cloudy

d. vibrio

e. sarcinae

f. rod

g. coccus

h. spirochete

i. staphylococci

j. concentric, contoured,
   smooth, radiated,
   wrinkled

_____ "comma"-shaped bacillus

_____ types of colonial edge characteristics on an agar plate

_____ description of the surface of bacterial colonies

_____ types of bacterial growth on an agar slant

_____ types of bacterial growth in broth

_____ rounded bacterial cell

_____ elongated bacillus

_____ corkscrew-shaped bacillus

_____ group of cocci arranged in packets of eight

_____ irregular groups of cocci with no specific arrangement or pattern

MULTIPLE CHOICE

1. "Hockey puck"-shaped bacterial colonies would be described as:

    a. raised   b. flat   c. even   d. radiated

2. Bacterial growth that completely covers the surface of a broth tube is termed:

    a. flat   b. pellicle   c. ring   d. meniscus

3. Tiny bacterial colonies growing on an agar plate may be termed:

    a. brush   b. pinpoint   c. stippled   d. fine

4. A broth tube with bacterial growth is properly mixed. Large masses of bacterial growth are seen suspended within the medium. This type of growth is termed:

    a. flocculation   b. ropy   c. pebbly   d. precipitation

5. A thin smear is the best way to prepare bacteria for viewing under the microscope because:

    a. the objective lens will not become contaminated
    b. a thick smear takes too long to stain
    c. it allows the viewer to properly observe bacterial group morphology
    d. it allows light to penetrate into the cells

6. The proper procedure for the simple staining of a bacterium is:

    a. air dry, add stain, rinse off, heat fix
    b. heat fix, rinse off, air dry, blot
    c. air dry, heat fix, stain, rinse off, blot
    d. heat fix, air dry, stain, rinse off

7. Which of the following is a description of the elevation of a bacterial colony above the surface of an agar plate?

    a. undulate   b. irregular   c. curled   d. convex

8. Which of the following is a description of the margin or edge characteristics of growth on a plate?

    a. serrated   b. curled   c. lobate   d. all of these

WORKING DEFINITIONS AND TERMS

Cellular morphology The shape and arrangement of cells as seen under the microscope.

Cultural morphology The appearance of bacterial growth as seen in broth cultures, on an agar slant, or on agar plates.

Simple stain A staining procedure in which all objects seen under the microscope are the same color. Different cellular structures may absorb different amounts of the stain, thus showing different shades of the stain.

EXERCISE 4

Bacterial Growth

Objectives

At the conclusion of this exercise students should be able to:

1. Define and describe bacterial growth.

2. Identify the components of the bacterial growth curve.

3. Determine viable cell numbers in a culture tube by performing a spread plate technique.

4. Quantitate bacterial numbers and determine a growth curve by taking turbidometric measurements of a culture by using a colorimeter.

Bacterial growth in the microbial world does not refer to cells that demonstrate continued increase in cell size. Instead, it refers to an increase in the cell number. Bacterial cells increase in size prior to **binary fission,** an asexual mechanism by which a cell gathers nutrients, duplicates its nucleic acids and proteins and splits into two daughter cells or clones itself. The time required for one cell to become two cells is referred to as the bacterial **generation time.** Each generation of cells is described by the equation 2^n where 2 identifies the number of cells that will be produced at a set generation time

and n is the number of cells: $2^5 = 32$ cells. Two is the number of cells formed with each generation; $5 =$ starting number of cells being considered. Therefore $2 \times 2 = 4 \times 2 = 8 \times 2 = 16 \times 2 = 32$. Since cell numbers in a broth culture environment often increase at rapid rates, bacterial numbers are expressed exponentially or logarithmically to the power of 10; that is, 500,000 cells/mL $= 5.0 \times 10^5$ cells/mL.

The growth phases exhibited by bacteria grown in broth cultures mimic the bacterial growth curve presented in Fig 4.1. The curve shows that there are four

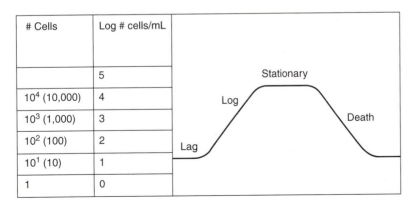

# Cells	Log # cells/mL
	5
10^4 (10,000)	4
10^3 (1,000)	3
10^2 (100)	2
10^1 (10)	1
1	0

FIG. 4.1. The bacterial growth curve.

phases of growth. Each phase has different events occurring within it which are described below.

1. *Lag phase:* The stage of bacterial growth where bacteria are acclimating to their new environment and are gathering nutrients in readiness for cell division, increasing their size, and synthesizing mainly enzymes.

2. *Log or exponential phase:* Here cells are duplicating at a constant rate and the cells are metabolically active. This is the stage when the generation time of a culture can be determined.

3. *Stationary phase:* A stage when cell death equals cell growth. At this point in the growth cycle there is no net increase in cell numbers. The medium at this stage contains limited nutrients and the presence of toxic waste products generated from metabolism in large amounts.

4. *Death phase:* This stage is marked by the accumulation of toxic substances which results in the decline of cell numbers. Many cells autolyse, and most cells have used up surrounding nutrients.

Growth of bacteria on solid agar media typically results in **Colony** formation. From one cell all cloned cells are developed resulting in evidence of visible growth. All phases of the growth curve occur within a colony. The outer peripheral edge has stopped growing because nutrients are minimal and a certain amount of cellular death is likely to have occurred.

FACTORS NEEDED FOR BACTERIAL GROWTH

Bacterial growth is influenced by physical and nutritional factors. The physical factors include *pH, temperature, moisture, oxygen concentration, and osmotic pressure.* The nutritional factors include the availability to cells of a *carbon source, nitrogen source,* presence in the growth environment of *sulfur, phosphorous, trace metals* and, in some instances, *vitamins.*

The optimal *pH* for most bacterial growth is approximately 7.0. Cells that grow with a pH of 5.4 to 8.5 are classified as Neutrophiles and most bacteria that cause human disease grow within this pH range. During growth, bacteria often produce metabolic waste products—either acids or bases that eventually interfere with their own growth. To prevent this situation, laboratory media often contain buffers such as phosphates to maintain the proper growth pH environment.

Most bacteria grow best at a *temperature* of 35–37° C. Human pathogens, bacteria that cause disease and infection in man, are mesophiles that have growth temperatures of 24 to 40° C. The temperature range over which an organism often grows is temperature at which its enzymes function maximally.

Bacteria that are actively metabolizing nutrients require a *water environment* for their survival. In fact most vegetative cells can survive for only a few hours without moisture. Cells that are sporeformers can have their spores remain in a dormant state in a dry environment for some time.

The *oxygen concentration* in the environment in which bacteria grow subdivides bacteria into two major categories. The aerobes which require O_2 in their environment for survival and the anaerobes which do not require oxygen in their environments. **Obligate aerobes** must have O_2 present whereas **obligate anaerobes** are killed with O_2. For aerobes, O_2 is needed for respiration and it is a limiting factor that will determine the rate of microbial growth. Microbes such as *E. coli,* used in this exercise and *Staphylococcus aureus* are **facultative anaerobes.** These organisms carry on aerobic metabolism when oxygen is present but shift to anaerobic metabolism when oxygen is absent. The facultative anaerobes have the most complex of enzymes systems since one set of enzymes enables them to use oxygen as an election acceptor and another set of enzymes is turned on when oxygen is not available.

Osmotic pressure is the regulation of water movement inside and outside of cells determined by the amount of dissolved substances found in the growth medium. Cells in a medium with high amounts of dissolved substances will lose water to their environments and shrink. Conversely, cells in water with nearly no dissolved substances will swell and burst. Most laboratory growth media contain the proper amounts of dissolved substances to allow cells to live and multiply without the constraints caused by osmotic pressure changes. Some bacteria such as *Staphylococcus aureus* are halophiles or salt-loving organisms desiring moderate to large amounts of salt (sodium chloride). These organisms have a cell wall and membrane that can tolerate high salt concentrations and have no interruption of metabolic function.

Nutritionally, bacteria require *carbon, nitrogen, sulfur, phosphorous,* and *trace elements.* Bacteria that have special nutrition needs are referred to as **fastidious.** These organisms require special additions to the growth medium such as various vitamins, blood or serum components for growth to occur. A *carbon source* is used by bacteria as a major metabolic energy source. Glucose is the most common carbon source and is metabolized by glycolysis, Krebs cycle, and/or by fermentation. The *nitrogen source* is usually provided by the addition of amino acids, peptides, petones, or nucleic acids. These biochemicals are used by bacteria as building blocks for its manufacture of bacterial proteins or DNA/RNA, respectively. Nitrogen can also be an additive to the growth medium as nitrate or ammonium ions. *Sulfur* is a mineral that is used by bacteria for the synthesis of sulfur-containing amino acids and for the manufacture of structural proteins. *Phosphorous* is a mineral that is used in the synthesis of phosphate ions that are used by cells to synthesize ATP, phospholipids and nucleic acids. The *trace elements* such as calcium, copper, zinc, iron, magnesium, and manganese are utilized as factors to activate enzymes, iron is specifically used to synthesize heme molecules and calcium is required by Gram (+) bacteria for cell wall synthesis and is used by spore-forming bacteria to manufacture spores.

Vitamins, if required by the bacteria for growth, are utilized by them for the manufacture of coenzyme molecules.

MEASURING BACTERIAL GROWTH

The number of cells that arise through binary fission can be measured by determining the viable cell number, which equals the number of living organisms/mL culture through either a pour plate or spread plate method. In the pour plate procedure a diluted bacterial culture is added to melted agar and this mixture is poured into an empty Petri dish. Once the plate cools it solidifies and it is then incubated at optimal temperature to develop colonies. Colonies in this method can develop on the surface and within the agar medium or can be heat damaged by the melted agar and never develop into colonies. In the spread plate method a tenth (0.1) mL of diluted bacterial suspension is applied to the center of an agar plate and it is spread out with the use of a curved glass rod. After incubation at the appropriate temperature the viable colony number is counted. Regardless of the viable cell method that is used, the countable number of colonies must average 30–300 colonies/plate. Duplicate plates at each dilution are performed so that an average colony number can be obtained. The number of colonies counted multiplied by the reciprocal of the dilution made = The Number of Bacteria/mL of original suspension. If 120 colonies were counted from a diluted suspension of 1/1000 (1:1000) then the bacteria/mL in the diluted suspension is $120 \times 1000 = 12.0 \times 10^4$.

I. Procedure for Preparation of Spread Plate:

Materials List Per Table/Workstation

24-hr culture *E. coli* in Brain Heart Infusion Broth

1 ml sterile pipettes

5 tubes of sterile H_2O (each tube contains 9 ml)

10 plates of Trypticase Soy Agar

5 bent glass rods (Hockey sticks)

Quebec colony counter

PROCEDURE

1. Inoculate *E. coli* into a Brain Heart Infusion Broth and incubate overnight at 37° C.

2. Prepare serial dilutions of the original culture tube by transferring 1 mL of culture into a 9-mL tube of sterile H_2O, mixing and removing from this dilution 1 ml to be transferred to another 9-mL sterile water blank tube. Refer to Figure 4.2 on following page. Prepare 5 *dilutions* (1:10, 1:100, 1:1,000, 1:10,000, 1:100,000) of the original bacterial suspension.

3. Dispense 0.1 mL of each dilution into two plates of Trypticase Soy Agar and spread each duplicate plate with the same bent glass rod. Turn each plate 45° and spread the diluted suspension on the agar surface in another direction in order to cover the total agar surface.

4. Invert all plates and incubate them at 37° C for 24 hrs.

5. Count those duplicate plates having 30–300 colonies each. Counting can be assisted by using a Quebec colony counter which is equipped with a magnifying lens and grid.

6. Calculate the average number of bacteria/mL by multiplying the average colony plate number by the reciprocal of the dilution plated. Record your results in Table 4.1.

Inventory

10 plates of Trypticase Soy Agar each labeled in duplicate with the 5 dilutions of bacterial suspension made.

II. Procedure for Demonstration of a Bacterial Growth Curve

Materials

24-hr culture *E. coli* in Brain Heart Infusion Broth

5 ml sterile pipettes

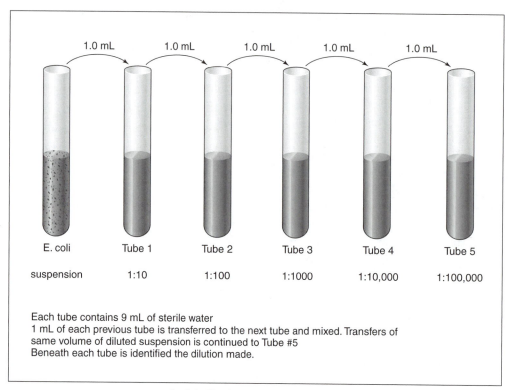

FIG. 4.2. Serial dilution of a bacterial suspension.

TABLE 4.1	RESULTS OF SPREAD PLATE TECHNIQUE FOR QUANTITATION OF VIABLE BACTERIAL NUMBERS/ML OF *E. COLI* SUSPENSION					
		Colony Counts*			Calculated**
Tube #	Serial Dilution	Plate 1	Plate 2	Average	Bacterial #/mL
1	1:10				
2	1:100				
3	1:1000				
4	1:10,000				
5	1:100,000				

* If colony counts are too high record TNTC—Too Numerous To Count
** Typical calculation. Average colony number counted = 12 for 1:1000; 12 × 1000 = 12,000 bacteria/ml in the original suspension.

4 13 × 100 mm glass tubes (used as cuvettes)

Parafilm squares

Colorimeter (spectronic 20) set at 600 nm

37° C incubator

PROCEDURE

In order to determine the bacterial growth number in a culture which includes viable and non viable cells the same *E. coli* culture grown overnight in Brain Heart Infusion Broth at an incubation temperature of 37° C in Part I will be used for this exercise.

A standard colorimeter is required for this exercise. It should be turned on for at least 15 minutes prior to use and adjust it to a wavelength of 600 nm.

Blank the colorimeter to a zero absorbance reading using a 13 × 100 mm tube containing 3 mL of Brain Heart Infusion Broth. As the culture grows in an inoculated tube of Brain Heart Infusion Broth there will be an increase in turbidity and a corresponding increase in Absorbance reading at 600 nm. Prior to each recorded absorbance reading make sure to suspend the culture tube so that the cells are distributed throughout the broth.

Follow the procedure below to semiquantitate the microbial numbers (growth) observed and to demonstrate a bacterial growth curve.

1. Dispense 3 mL of sterile Brain Heart Infusion Broth into 4 13 × 100 mm tubes which will be used as cuvettes to monitor bacterial growth. Cover each tube with parafilm to prevent contamination.

   Tube #1 will be used to calibrate the colorimeter and to adjust the absorbance to zero at 600 nm.

   Tube #2 add 0.1 mL of overnight *E. coli* culture

   Tube #3 add 0.2 mL of overnight *E. coli* culture

   Tube #4 add 0.5 mL of overnight *E. coli* culture

2. Cover each tube with parafilm, invert and take an initial absorbance reading and record this value in Table 4.2, then place the tube at 37° C for 20 minutes.

3. Remove the tube from the incubator at 20 minutes and invert the tube and read the absorbance at 600 nm. Repeat at 20-minute intervals for 3 hours, recording the absorbance reading obtained of each tube and record your results in Table 4.2.

4. Plot the absorbance obtained for each tube versus the time the reading was made in Figure 4.3.

5. Describe the bacterial growth curve from the absorbance values obtained.

TABLE 4.2

ABSORBANCE VALUES OF EACH TUBE

Time (minutes)	Absorbance @ 600 nm Tube # (dilution)		
	#2 (¹⁄₁₀)	#3 (²⁄₁₀)	#4 (⁵⁄₁₀)
0			
20			
40			
60			
80			
100			
120			
140			
160			
180			

Plot the absorbance of each tube versus the time reading was made on Figure 4.3.

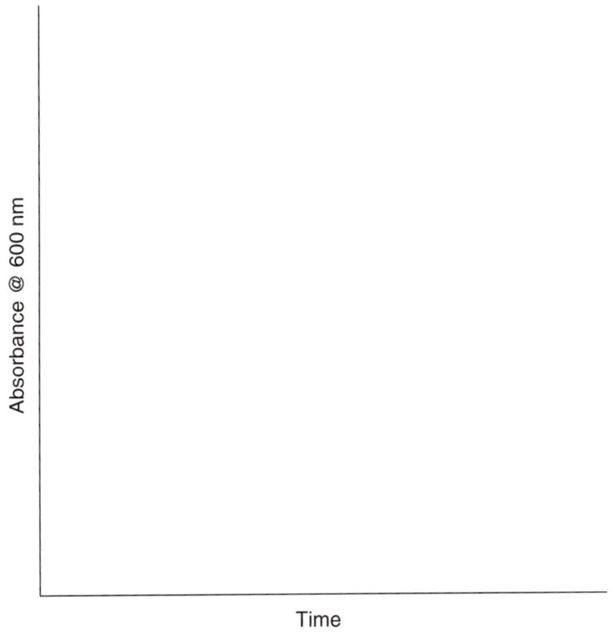

FIG. 4.3. Bacterial growth curve.

NAME _____ DATE _____ SECTION _____

CRITICAL THINKING QUESTIONS

1. What factors may be the cause for an extended Lag phase in a bacteria growth curve?

2. If you altered the conditions under which bacterial growth normally occurs (i.e.) increase the temperature of incubating a culture from 35° to 40° C, what effect would this have on the bacterial growth curve of the organism under study?

3. How might the bacterial growth curve change if a facultative anaerobe was first monitored for growth when grown in the presence of an O_2 environment and during its log (exponential) growth phase the organism was suddenly placed in an anaerobic environment?

4. A bacterial growth curve was determined in this exercise for *E. coli* by measuring the absorbance of the culture at 20-minute intervals. When the curve showed plateaus (Lag and Stationary Phases) was the absorbance obtained proportional to the number of bacteria present in the cuvette tube?

FILL-IN QUESTIONS

1. The phase of the bacterial growth curve where cells are dying at rapid rates is called the _____ phase.

2. Bacterial cells are most likely to be affected by antibiotics when they are in this phase of the bacterial growth curve _____.

3. If one ml of a diluted culture (1:1000) was added to 9 ml of water, the dilution made is _____.

4. An organism that strictly requires an atmosphere of CO_2 for growth is called a _____.

5. List five physical factors required by bacteria for growth: _____, _____, _____, _____.

6. Identify three nutritional factors required by most bacteria for growth: _____, _____, _____.

7. A microbe that requires serum components to be added to the growth medium is called _____.

8. What is a countable number of colonies on a bacterial plate? _____

9. A dilution made whereby the liquid transferred from tube to tube is uniform and the total volume in each tube remains uniform is called _____.

10. Measurement of a bacterial growth curve is referred to as a _____ (quantitative or semi-quantitative) procedure.

MATCHING QUESTION

You may use each growth phase more than once to match with the characteristics of growth identified.

Characteristic of Growth

_____ 1. No net increase in cells has occurred.

_____ 2. Cells experience autolysis.

_____ 3. Cells acclimate to the environment.

_____ 4. Cellular generations have occurred in culture.

_____ 5. Binary fission is maximal.

_____ 6. Cells are producing toxic byproducts of metabolism.

_____ 7. Cells that are sporeformers will generate spores at this phase.

_____ 8. Cells are gathering nutrients and synthesizing macromolecules for cell division.

_____ 9. No increase in absorbance occurs at these growth phases.

_____ 10. Antibiotics probably have their greatest killing effectiveness at this phase of growth.

Growth Phase

a. Lag

b. Log/Exponential

c. Stationary

d. Death

WORKING DEFINITIONS AND TERMS

Bacterial growth A series of stages exhibited by bacterial cells characterized by an increase in cell numbers.

Binary fission A bacterial cell that divides or splits into two equal size cells. This is the mode of asexual reproduction in bacteria.

Colony Isolated visual growth on solid agar plate of bacteria that have cloned themselves.

Colorimeter An instrument designed to measure the absorbance of solutions using the visible spectrum of the light range (340 nm to 700 nm).

Facultative anaerobe Organisms which can grow in the presence and absence of O_2.

Fastidious Organisms which have special nutritional needs for growth to occur.

Generation time The amount of time it takes for actively dividing bacterial cells to produce two cells from one original cell.

Obligate aerobe A strict aerobe which requires O_2 for growth.

Obligate anaerobe A strict anaerobe requiring an atmosphere without O_2 for growth.

Serial dilution A uniform dilution made by distributing a set amount of a solution into a tube, mixing the tube and transferring the same amount of solution to a subsequent tube filled with the same amount of diluent.

Spread plate A method used to quantitate bacterial numbers from a diluted suspension of cells applied to an agar plate.

Turbidimetry Measuring the amount of turbidity exhibited by a solution that is suspended and is placed in a colorimeter.

Gram Stain and Acid-Fast Stain

Objectives

After completing this lab, you should be able to:

1. Describe the principle of the differential stain.

2. Properly perform a Gram stain.

3. Differentiate between Gram positive, Gram negative, and Gram variable reactions.

4. Properly perform an acid-fast stain.

5. Differentiate between an acid-fast and a nonacid-fast staining reaction.

You were introduced to the simple stain in Exercise 3, where all the microbes seen with the microscope were the color of the single stain used for the preparation. *Differential staining* requires more than one type of stain and is used to distinguish between various types of bacterial cells. A differential stain typically consists of three main steps: first, a *primary stain,* which is used to stain all the cells on the slide; then a *decolorizing* step, which removes the stain only from certain types of cells; and finally, a *counterstain,* which stains the newly decolorized cells but has no effect on the cells still holding the primary stain. The Gram stain and acid-fast stain are two widely used differential stains used to distinguish and classify bacteria according to their cell walls.

THE GRAM STAIN

Hans Christian Gram (1853–1938) discovered and perfected the Gram stain in the 1880s while working on a technique to detect mammalian cells infected with bacteria. His discovery was soon used to divide bacteria into two main groups—Gram positive and Gram negative—as well as two smaller groups—Gram nonreactive and Gram variable. The Gram stain reaction is based on the amount of peptidoglycan found in the cell walls of these bacteria. Gram positive bacteria have many layers of peptidoglycan, which, in turn, holds molecules of teichoic acids. Gram negative bacteria have only one layer of peptidoglycan with no teichoic acid. Teichoic acid reacts with the crystal violet and iodine used in this staining process. A complex of crystal violet–iodine–teichoic acid molecules form, which results in large, difficult-to-remove complexes. Since Gram positive cell walls hold many of these complexes, it is more difficult (but not impossible) to decolorize a Gram positive cell than a Gram negative one. An alcohol mixture readily removes the crystal violet from the Gram negative cell but not from the Gram positive. This alcohol mixture also dissolves much of the lipopolysaccharide outer layer of the Gram negative cell wall, which further speeds the removal of the crystal violet primary stain from these cells.

When another stain, usually safranin, is added, the Gram positive cells, still stained with the much darker crystal violet–iodine complex, will not show this lighter stain. The now colorless Gram negative cells will soon absorb the pinkish red color of the safranin. At the conclusion of the Gram stain procedure, the Gram positive cells will be the color of crystal violet, or the primary stain, and the Gram negative cells will be the color of safranin, which is the counterstain.

Some bacterial cells are made up of thick, heavy lipids, which make them "waxy" and thus nearly waterproof. If water cannot penetrate, neither can the dyes dissolved in the water. *Mycobacteria*, the agents that cause tuberculosis and Hansen's disease (leprosy), are examples of such bacteria and are considered to be Gram nonreactive. A special staining procedure, called the acid-fast stain, is used to colorize such cells and will be covered later in this laboratory exercise.

Finally, some bacteria are considered Gram variable. That is, some cells retain the crystal violet stain, while others display the color of the counterstain, safranin. Four factors determine whether a cell is Gram variable.

1. *Genetics.* Some cells allow variable amounts of teichoic acid to build up in the cell wall, causing a variable reaction.

2. *Age of culture.* The Gram stain should be performed on a fresh, 18- to 24-hour culture. Older cultures develop variable amounts of teichoic acid in the cell wall, which causes variations in the iodine–crystal violet–teichoic acid reaction, which, in turn, causes a variable reaction.

3. *Type of growth medium.* Certain types of growth media do not contain the nutrients necessary for normal cell wall development. (You will not be using such media during the first part of this course.)

4. *Your technique.* If the smear is not thinly or evenly made, or if the staining procedure is not performed correctly, the cells will appear Gram variable.

The Gram Stain Technique (Traditional Method)

Microbes per table/workstation:

Corynebacterium xerosis, Escherichia coli, Micrococcus luteus, or *Sarcina flava*

Materials

Staining tray

Marking pen or pencil

Crystal violet

Gram's iodine

Gram's decolorizer (alcohol mixture)

Safranin

Bibulous paper

Glass slides

PROCEDURE

1. Prepare smears of *C. xerosis, E. coli,* and *M. luteus/ S. flava.* Air dry and heat fix. (See Exercise 3 for review of smear preparation.)

2. *Primary stain.* Place the slide on the staining tray and cover smears with crystal violet for approximately 1 minute.

3. Rinse slide with water.

4. *Mordant or fixative.* Cover the smear with Gram's iodine, rotate and tilt the slide to allow the iodine to drain, and then cover again with iodine for 1 minute. Since iodine does not mix well with water, this procedure ensures that the iodine contacts the cell walls of the bacteria on the slide.

5. Rinse slide with water as in step 3.

6. *Decolorize.* Place several drops of Gram's decolorizer (alcohol) evenly over the smears, rotate, and tilt the slide. Continue to add alcohol until *most* of the excess stain is removed and the alcohol running from the slide appears clear. This is the most critical step of the procedure. If the smear is too thick, or if the alcohol is kept on the slide too long or too short a time, the results will not be accurate. Although there is no recommended time for this step, it usually takes between 5 and 15 seconds to decolorize a thin smear properly.

⚠ **REMEMBER EVEN GRAM POSITIVE CELLS WILL DECOLORIZE IF EXPOSED TO THE DECOLORIZER LONG ENOUGH!**

7. *Immediately* rinse off with water.

8. *Counterstain.* Add safranin solution for approximately 30 seconds. Colorless Gram negative cells will readily accept the light red safranin stain, while the already dark-colored Gram positive cells will undergo no color change at all.

9. Rinse off with water, and blot dry with Bibulous paper or a paper towel.

10. Pour the residual stain in the staining tray into an appropriate discard container.

Gram Stain: (Alternative Method)

Recently a new technique for performing the Gram stain was developed. In this new method, the decolorizing and counterstain steps are combined by using a modified safranin stain.

PROCEDURE:

Perform steps 1-4 as described above in the traditional method.

5. Wash off the iodine solution using the safranin-decolorizer mixture.

6. Immediately reapply the safranin-decolorizer solution and let it sit for 20-50 seconds.

7. Rinse and blot dry.

There is still another method to identify Gram positive and Gram negative cell walls. This method can be utilized on certain Gram positive bacteria that are very easily decolorized and thus appear to be Gram negative. Potassium hydroxide (KOH) is used in this technique. The procedure involves placing 2 drops of a 3% KOH solution on a slide and mixing for 30 seconds with a loopful of the test organism taken from a pure colony. If the bacteria is Gram negative, the KOH will break down its cell walls, causing chromosomal DNA to be released, which will then become stringy. This stringiness can be detected by periodically lifting up the inoculating loop while mixing the bacteria in the KOH solution. Gram positive organisms will not form these strings. (See Fig. 5.1.)

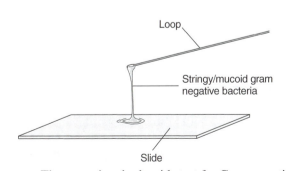

FIG. 5.1. The potassium hydroxide test for Gram negative bacteria.

Results

TABLE 5.1 GRAM STAIN REACTIONS		
Microbe	*Staining Reaction*	*Morphology*
Corynebacterium xerosis		
Escherichia coli		
Micrococcus luteus or *Sarcina flava*		

THE ACID-FAST STAIN TECHNIQUE

As mentioned previously, certain bacterial cell walls contain high concentrations of dense "waxy" lipids that prevent the penetration of water. If water cannot enter these cells under normal circumstances, neither can any dye dissolved in the water. The acid-fast stain uses a procedure that forces dye through this nearly waterproof cell wall. Once inside, the dye is virtually trapped inside and even resists decolorization with an *acid-based decolorizer. Color fastness* is a characteristic of certain microbes that resist decolorization with *acid alcohol.* Normal vegetative cells are almost immediately decolorized with such an alcohol solution. Thus, when a counterstain is added to the slide, these decolorized cells readily absorb the new dye.

A century ago, Paul Ehrlich discovered this procedure while working with Robert Koch on the problem of staining *Mycobacterium tuberculosis,* the causative agent of tuberculosis. One of the virulence factors of this microbe is its extremely thick, waxy cell wall. It protects the microbe from many disinfectants, and from drying out; this cell wall even protects the microbe from our immune system. Hospitals and clinical facilities that process suspected tuberculosis specimens use separate rooms or specially designed transfer hoods for transferring and staining these dangerous microbes. Because it would be extremely hazardous (and illegal) to use such a bacterial cell for this exercise, a cell with a much thinner cell wall will be substituted. Therefore, you should find that it will be very easy to over-decolorize these slides.

Acid-Fast Technique

Materials List Per Student/Workstation

Microbes per table/workstation: *Mycobacterium smegmatis*

Carbol Fuchsin stain

Acid alcohol decolorizer

Methylene Blue Stain or Brilliant Green Stain

TRADITIONAL PROCEDURE

The traditional method uses heat and time to allow a lipid-penetrating dye to enter the nearly waterproof cell wall of acid-fast bacteria.

1. Prepare smears of *Mycobacterium smegmatis* and *Micrococcus luteus* or *Sarcina flava* on the same slide. Allow to air dry and heat fix. If directed to do so by your lab instructor, cover the slide with a rectangular piece of paper towel.

2a. Place some water on the bottom of your staining tray, place the slide on the tray, and flood the slide with carbol fuchsin stain. Make sure the entire slide is covered. Maneuver the Bunsen burner so that the flame heats the smear *from underneath the slide.* Continue heating until the slide begins to steam. Once the slide stops steaming, reheat. (An alternative method is to use a hot plate.) (See Fig. 5.2.)

   OR

2b. Place the slide on a wire mesh placed over a beaker of water, which in turn is placed on a heating tripod. Flood the slide with carbol fuchsin stain. Make sure the entire slide is covered. Heat the water to boiling and allow the slide to remain over the boiling water for 5 minutes. (See Fig. 5.3.)

3. If the dye evaporates or runs off the slide, add more dye. If a paper towel rectangle was placed on the slide, remove it, discarding it in the waste basket. *Do not allow the stain to dry out.*

4. Rinse off with water. If stain adheres to the bottom of the slide, gently rub it off with a piece of paper towel saturated with acid alcohol.

5. Decolorize by covering the slide with the acid alcohol solution until the alcohol runs clear. Then *immediately* stop the reaction by flooding with water.

6. Counterstain with methylene blue for 1 minute, rinse off, and blot dry. (Your instructor may direct you to use brilliant green as a counterstain.)

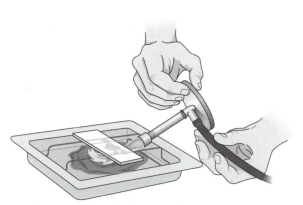

FIG. 5.2. Heat the slide from below.

FIG. 5.3. Slide on beaker. Heat acid-fast slide over a beaker of boiling water.

7. Discard excess stain as directed.

ALTERNATIVE PROCEDURE

A variation of the traditional acid-fast stain is one that uses no heat. This eliminates the possibility of burnt fingers during the heating process; it also prevents various chemical fumes from wafting throughout the laboratory.

After preparing the smear, flood the slide with the modified Kinyoun carbol fuchsin stain for 2 to 5 minutes. *Do not allow the stain to dry out!* Rinse off with water and complete steps 5–7 above.

Results

TABLE 5.2	ACID-FAST STAINING REACTIONS		
	Microbe	Staining Reaction	Morphology
	M. smegmatis		
	M. luteus or S. flava		

Inventory

After completing this exercise, you will have done the following:

A Gram stain slide showing a Gram positive, Gram negative, and possibly a Gram variable reaction.

An acid-fast slide showing an acid-fast and a nonacid-fast reaction.

Any practice tubes and streak plates to continue transfer skills.

LABORATORY CLEANUP

Incubation

Incubate any practice tubes and transfer plates as described in Exercise 2.

Discards

Discard all tubes and plates not placed in the incubation tray as described in Exercise 2.

General Cleanup

Clean your tabletop and work area, store your lab coat, and put away all equipment as described in Exercises 1 and 2.

Slides and Microscopes

Store or discard your prepared slides and clean your microscopes as directed in Exercises 1 and 3.

Note: Future exercises will not include a laboratory cleanup section within the text. In future exercises, refer to Exercise 2 for a review of laboratory cleanup procedures.

NAME _____ DATE _____ SECTION _____

QUESTIONS

1. What are the four possible results of a Gram stain?

2. Which stain is the primary stain for the Gram stain, and which one is the primary stain for the acid-fast stain?

3. Which stain is the counterstain for the Gram stain, and which stain is the counterstain for the acid-fast stain?

4. What are some of the reasons for a Gram variable reaction?

5. The acid-fast bacterium, *Mycobacterium smegmatis,* is relatively safe for students to work with because of its thin cell wall. If a Gram stain is to be performed on this particular microbe, it often takes the Gram stain reaction. Why?

MATCHING

a. green

b. blue

c. red

d. violet

e. safranin

f. alcohol

g. simple

h. differential

i acid-alcohol

_____ color of Gram negative staining reaction under the microscope

_____ chemical used as a decolorizer in the acid-fast stain

_____ color of Gram positive staining reaction under the microscope

_____ chemical used to decolorize the Gram stain

_____ type of stain where various types of microbes can be identified based on color

_____ color of acid-fast staining reaction under the microscope

(Answers may be used more than once.)

MULTIPLE CHOICE

1. The stain used to visualize *Mycobacteria* is:

   a. simple stain   b. Gram stain   c. acid-fast stain   d. endospore stain

2. What color would Gram positive cells show under the microscope?

   a. pink   b. violet   c. blue   d. green

3. A normally Gram positive cell shows up as Gram variable. Which explanation reveals why this occurred?

   a. the smear was too thick
   b. the smear was exposed to the decolorizer too long
   c. a 48-hour culture was used
   d. all of these

4. A microbial cell that shows up positive in a differential stain:

   a. is impermeable to water
   b. has a thick cell wall
   c. has been decolorized
   d. retains the color of the primary stain

5. The chemical found in Gram positive cell walls and not in Gram negative cell walls that reacts with the crystal violet–iodine complex is:

   a. lipopolysaccharides   b. teichoic acid   c. protein coat   d. peptidoglycan

6. The iodine in the Gram stain is used:

   a. to remove safranin from the cell wall
   b. to decolorize
   c. as a primary stain
   d. as a fixative or mordant

7. Which stain is used as the counterstain for the Gram stain?

   a. iodine   b. safranin   c. crystal violet   d. methylene blue

WORKING DEFINITIONS AND TERMS

Acid-fast organism Any microbe that resists decolorization with an acid-alcohol solution.

Counterstain The second stain used in differential stains. Cells that display this color are usually considered "negative."

Decolorizer A chemical used to remove the primary stain from some bacterial cells during a differential staining reaction. The microbes that become decolorized are "negative," and the microbes that retain the primary stain are "positive."

Differential stain Any staining technique that categorizes cells based on how they react to dyes present in the stains.

Gram stained bacterium Any bacterial cell that accepts the Gram stain reaction

Primary stain The initial stain used in a differential stain. Microbes that retain this stain after the process is complete are usually considered "positive."

Endospore Stain, Capsule Stain, and the Hanging Drop Technique

Objectives

After completing this lab, you should be able to:

1. Properly perform an endospore stain.

2. Differentiate between endospore formation and nonendospore formation.

3. Recognize the presence of bacterial capsules under the microscope.

4. Properly perform a hanging drop technique for detecting bacterial motility.

5. Recognize the difference between a motile and nonmotile bacterium based on the hanging drop technique.

You have already been introduced to the differential stains used to classify bacteria by the chemical composition of their cell walls. There are also differential stains that allow you to recognize various cellular types based on their structures; endospores, capsules, and flagella are examples of such structures. In this exercise, you will perform the endospore stain and the capsule stain. Although there is a staining procedure for detecting the presence of flagella on bacteria, it is rarely done because it is too difficult. Another way of detecting the presence of flagella on bacteria is to look for bacterial movement or motility under the microscope in a procedure known as the *hanging drop technique.*

THE ENDOSPORE STAIN

Spore-forming bacteria are responsible for several serious diseases as well as one type of food poisoning. The *Clostridia,* a genus of anaerobic bacteria, contains the species responsible for gas gangrene, botulism, and tetanus. The genus *Bacillus* includes a species that causes the disease anthrax. Stepping on a rusty nail has

long been associated with the disease tetanus, but it is the spores on the nail and not the rust that is responsible for this disease. Since bacteria form their spores within their vegetative cells, they are called *endospores.*

Bacterial endospores are made up of genetic material, heat-resistant enzymes, very little water, and a thick, waterproof, outer protein called the *spore coat.* This spore coat performs a similar function for the spore as the lipid-rich cell wall does for the acid-fast bacteria (AFB), which prevent the penetration of water. This coat also makes the spore highly heat and disinfectant resistant. A staining process similar to the acid-fast staining procedure is used to force dye through the spore coat. The dye, malachite green, acts as the primary stain and will color everything green. Although malachite green is somewhat soluble in water, it will be too time consuming to use only water to act as a decolorizing agent. Safranin, the counterstain used in the Gram stain procedure, will not only act as the counterstain here, but will also replace the malachite green in the vegetative bacterial cells. Since the safranin cannot penetrate the spore coat, the spores remain green (primary stain) and the vegetative cells show up as the color of safranin (counterstain). Only the *Bacillus* and *Clostridium spp.*

are known as endospore formers, and these are both rod-shaped bacteria.

Since spores are produced from vegetative cells, you can expect to see rod-shaped, safranin-stained vegetative cells mixed in with the spores. However, *everything green under the microscope is not a spore.* Certain vegetative cells, for example, *Mycobacteria,* dust, and other debris often are not decolorized and will appear green under the microscope. Therefore, you have to be able to recognize the characteristic shape of the spore as well as its color. Although the endospore is produced within the bacterial cell, once it is fully formed, the vegetative cell that produced it, dies. This dead cell is rather brittle and easily breaks apart when placed on a smear. Expect to see many "free" spores or "exospores" under the microscope rather than spores within cells.

Outside all bacterial cells with cell walls there is a layer of polysaccharides known as glycocalyx. If this layer is thick, organized, and densely packed, it is called a *capsule.* Many bacterial cells that have capsules are virulent pathogens or have the ability to easily cause disease in healthy people. The causative agent of the most common form of bacterial pneumonia, *Streptococcus pneumoniae,* owes its virulence to this capsule, for forms of this microbe without the capsule are relatively harmless. Capsules prevent bacterial cells from drying out when they are on environmental surfaces, protect from disinfectants, as well as ingestion and digestion from white blood cells.

Certain bacterial cells are also capable of producing flagella, long whiplike structures that allow them to move. This *motility* enables them to travel on their own and thus acts as a spreading factor. For example, *E. coli* is a leading cause of urinary tract infection in catheterized patients in hospitals. One major way this microbe gets into the urinary bladder is to swim "upstream" through the catheter tube and into the bladder.

The Spore Stain Technique

Materials List Per Table/Workstation

Bacillus cereus, Bacillus subtilis, Staphylococcus aureus

Malachite green, 5% or 10% concentration

Safranin

Glass slides

Staining tray

TRADITIONAL PROCEDURE

As is true of the traditional acid-fast stain, this procedure uses heat and time to force the primary stain, malachite green, through the waterproof spore coat.

1. Prepare smears of the three assigned bacteria, air dry, and heat fix. If directed to do so by your lab instructor, place a rectangle made from a paper towel on top of the slide.

2. Flood the slide with 5% malachite green and heat using the same procedure as for the acid-fast staining procedure. Apply the flame from beneath the slide and heat until steaming. Heat for 5 minutes. Continue to add dye as needed. *Do not allow the dye to dry out!* After 5 minutes, allow the slide to cool. If a paper towel rectangle was placed on the slide, remove it and discard it in the wastepaper basket.

3. Rinse off excess dye with water.

4. Decolorize/counterstain with safranin for 1 minute. The safranin stain penetrates the vegetative cells and removes the malachite green, but it cannot do so to the spores.

5. Rinse off, blot dry, and observe under the microscope.

6. Pour the residual stain in the staining tray into an appropriate discard container.

ALTERNATIVE PROCEDURE

As with the alternative acid-fast procedure in the previous exercise, the spore stain also uses a modification of the malachite green formula to stain bacterial endospores without using heat. With this modification, a higher concentration of 10% malachite green is used. Perform this stain as above but without the use of heat. *Do not allow the dye to dry out!* Remember, do not use heat. Once this process is complete, process the slide following steps 3 to 6 above.

Results

TABLE 6.1	Microbe	Reaction	Morphology
	Bacillus cereus		
	Bacillus subtilis		
	Staphylococcus aureus		

THE CAPSULE STAIN

The unique feature of the capsule stain is that everything in the field of vision under the microscope becomes stained *except the capsule*. The capsule stain is, therefore, considered to be a *negative stain*. When completed, the background will be a dark color, the vegetative cells will be the color of safranin, and the capsules will be colorless. India ink is used to color the background. Since India ink is a very coarse dye, it cannot penetrate the capsule, and it settles around the outside of cells. This is what gives the background its dark color. Safranin penetrates the capsule and stains the cell, but does not adhere to the capsule. Once completed, the only microbial structure under the microscope that is not stained will be the capsule. Smears stained for capsule observation are not heat fixed since heating may alter the appearance of the capsule.

The Capsule Stain Technique

Materials List Per Table/Workstation

> *Klebsiella pneumoniae*
> India ink
> Safranin
> Glass slides

PROCEDURE

1. Place a small drop of India ink at the end of one slide.

2. Aseptically mix in a small amount of bacteria with your loop. Use the same technique for preparing a smear but do not spread the bacteria around. Flame the loop and return it to the test-tube rack.

3. Place the edge of a second slide at a 45 degree angle across the bacteria–India ink mixture. Allow the mixture to spread across the width of the slide.

4. Push (or pull) the second slide across the length of the first slide, which will spread the bacteria–India ink mixture evenly over the first slide. (See Fig. 6.1.) *Discard the second slide in disinfectant.*

5. Allow the slide to air dry but *do not heat fix.* Heat fixing causes the capsules to shrink.

6. Cover the slide with safranin for 30 seconds and gently rinse off.

7. Do not blot dry, for the slide was not heat fixed. Tilt the slide on its side to allow water to drain off. Air dry. Observe under the microscope.

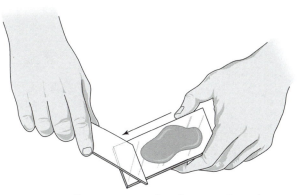

FIG. 6.1. Smear preparation for capsule stain.

Results

	Microbe	Reaction-Appearance	Morphology
TABLE 6.2	*Klebsiella pneumoniae*		

The Hanging Drop Technique

Materials List Per Table/Workstation

> Broth cultures of *Proteus mirabilis, Bacillus cereus* or *subtilis*
> Depression slides
> Cover slips
> Toothpicks or wooden stirring rods
> Petroleum jelly

PROCEDURE

1. Surround the well of the depression slide with a *thin layer* of petroleum jelly.

2. Aseptically place 1 to 2 loopfuls of a well-mixed broth solution of the test bacteria in the center of a cover slip. Do not spread it out.

3. Press the inverted depression slide onto the cover slip to seal the sample of broth within the petroleum jelly and quickly turn it right side up. (See Fig. 6.2.)

4. Let the slide sit a few minutes to allow the mixture to settle. (This prevents confusion as initially non-motile bacteria will also appear to have flagella.)

5. Carefully place the slide under the microscope and observe the drop under high power or oil immersion. Take care not to suddenly move or jar the slide. (Why?)

6. Look for bacteria coming in and out of the field of vision. *Hint:* Reducing the light and observing the

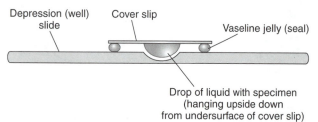

Depression (well) slide   Cover slip   Vaseline jelly (seal)

Drop of liquid with specimen
(hanging upside down
from undersurface of cover slip)

FIG. 6.2. The hanging drop technique.

edge of the hanging drop usually makes it easier to determine motility.

7. When completed, either discard the slide or recycle as directed.

Results

Determine which one of the two microbes tested was motile and which one was nonmotile.

NAME _____ DATE _____ SECTION _____

QUESTIONS

1. In what way is the endospore procedure similar to and dissimilar from the acid-fast staining procedure?

2. Sometimes acid-fast bacteria accept the spore stain and spores accept the acid-fast stain. Why?

3. Comment on the following statement regarding the spore stain procedure: "Everything green under the micro-scope is a spore."

4. During the preparation of the capsule stain, the slide used to spread the India ink/bacteria mixture is discarded in the disinfectant solution. Why?

5. What are some of the ways in which you can differentiate spores from coccus-shaped bacteria?

6. A hanging drop technique procedure was performed on a known motile bacterium. The slide was left on the microscope with the light on for several minutes. When finally observed, motility was not seen. What could be a possible explanation?

REVIEW OF STAINING PROCEDURES AND MORPHOLOGY

MATCHING

a. endospores

b. India ink

c. safranin

d. malachite green

e. carbol fuchsin

f. methylene blue

g. crystal violet

h. iodine

i. acid-fast stain

j. Gram positive

k. Gram negative

_____ used as a fixative in the Gram stain

_____ primary stain in the acid-fast staining procedure

_____ used to color the background in the capsule stain

_____ used to color the vegetative cells in the spore stain

_____ used to color the cells in the capsule stain

_____ used to color the endospores in the spore stain

_____ retains the crystal violet stain and is not decolorized by plain alcohol

MICROSCOPIC MORPHOLOGY

a. staphylococcus

b. sarcinae

c. single rod

d. streptobacillus

e. snapping

f. diplococci

g. streptococci

h. vibrio

i. spirochete

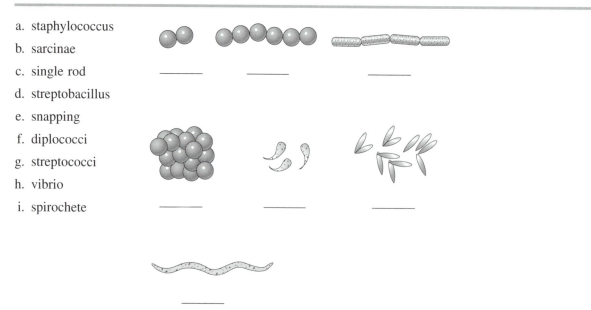

MICROBIAL MORPHOLOGY, DIFFERENTIAL STAINS

MULTIPLE CHOICE

1. Which of the following would you expect to see in a positive spore stain?

    a. green endospores
    b. pink vegetative cells
    c. green exospores or free spores
    d. all of these

2. What is the decolorizer for the spore stain?

    a. water   b. safranin   c. Gram's decolorizer   d. 3% acid alcohol

3. What is the color of the vegetative cell in a capsule stain?

    a. blue   b. pink   c. green   d. none of these

4. What is the color of the capsule in the capsule stain?

    a. blue   b. pink   c. green   d. none of these

5. The time needed for malachite green to penetrate spores is:

    a. 5 minutes   b. 1 minute   c. 30 seconds   d. 5 seconds

6. The counterstain for the Gram stain reaction is:

    a. safranin   b. methylene blue   c. malachite green   d. brilliant green

7. The primary stain for the acid-fast staining procedure is:

    a. methylene blue   b. brilliant green   c. malachite green   d. carbol fuchsin

8. A Gram nonreactive bacterium would most likely be:

    a. capsule forming   b. Acid-Fast   c. spore forming   d. a vibrio

9. In the staining procedure used to identify *Mycobacterium tuberculosis*, the decolorizer used is:

    a. acid alcohol   b. Gram's decolorizer   c. iodine   d. safranin

10: India ink is used in the _____ stain.

    a. Gram   b. Acid-Fast   c. spore   d. capsule

11. Bacterial motility can be detected by using:

    a. Gram stain
    b. acid-fast stain
    c. capsule stain
    d. hanging drop procedure

WORKING DEFINITIONS AND TERMS

Endospore A dormant form of a bacterium that is able to resist harsh environmental conditions.

Hanging drop procedure Nonstaining procedure used to detect the presence of flagella.

Negative stain Staining procedure whereby everything but the structure to be observed is stained, leaving the target structure colorless.

Vegetative cell An actively growing (bacterial) cell.

EXERCISE 7 Fiongi

Objectives

After completing this lab, you should be able to:

1. Properly differentiate between the four major divisions of fungi.

2. Identify sporangiophores, sporangiospores, and sporangia.

3. Identify zygospores.

4. Identify conidiophores, and conidia.

5. Differentiate between septate and nonseptate hyphae.

6. Identify a mycelium, vegetative and aerial hyphae, and rhizoids.

7. Identify yeast cells and budding.

THE FUNGI

Members of the kingdom Fungi are eukaryotic, plantlike organisms that possess a cell wall but cannot photosynthesize. The study of this kingdom is called *mycology.* All mycota are *saprophytes,* organisms that feed on dead or decaying matter. All mycota are acidophiles, favoring low pH. The kingdom contains both multicellular organisms, known as molds, and unicellular organisms, known as yeasts.

The most common form of the fungi are the molds. Some, but not all, molds can produce the yeast forms, and these molds are referred to as *dimorphic* (two shapes).

Molds grow as hairlike filaments called *hyphae.* The hyphae may or may not contain internal cross walls or *septa.* Hyphae with septa are termed *septate,* and hyphae without septa are *nonseptate.* (See Fig. 7.1.) When individual hypha grow together into a mass of tangled hyphae the mass is called a *mycelium,* and it is this mycelial mass that is familiar to us as a "mold."*

The vegetative hyphae of the mycelium easily grow on the surface of a piece of fruit or agar plate. The vegetative hyphae send very fine filaments, called *rhizoids* (*rhizo* = root) below the surface in a rootlike system to absorb nutrients.

Molds reproduce by forming asexual or sexual spores. These spores become easily airborne to facilitate their dispersal and reproduction. Many of these airborne spores are the cause of allergies and act as notorious sources of contamination in microbiology laboratories.

Yeasts are unicellular fungi that are oval in shape. Because they are much larger than bacterial cells, they are easily differentiated from bacterial cocci by their size. Yeast cells, unlike bacterial cells, possess a nucleus. Yeasts reproduce asexually by budding. A *bud* is a tiny, oval-shaped extension from the parent cell. The bud en-

---

*So you like blue cheese and some crusty bread, along with a nice glass of wine. Thank the mold. Strains of *Penicillium roquetforti* are responsible for the taste of blue cheese and Gorgonzola. (Guess what the blue-green flecks are?) *Saccharomyces cerevisiae* is the most common form of yeast used to make bread rise. Although alcohol is a byproduct of the fermentation of sugar, it is driven off in the baking process. If you like San Francisco sourdough bread with your cheese, the bacterium *Lactobacillus sanfrancisco* also becomes part of the recipe. When *Saccharomyces cerevisiae* is added to grape juice with the proper concentration of glucose or fructose (technically called "must" in the wine industry), you will eventually have some wine with that cheese and bread.

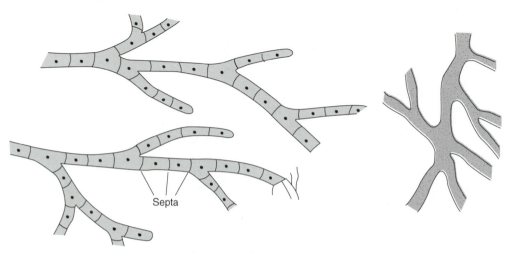

FIG. 7.1. *Septate and Non-septate Hyphae*

larges in size and eventually pinches off of the parent cell. A common yeast is *Saccharomyces cerevisiae* (sugar fungus), a member of the Ascomycota division. This is the yeast associated with making bread rise and with fermenting wine and beer. Other yeasts are normal inhabitants or *flora* of our mouth, skin, and colon. *Candida albicans,* a member of the Deuteromycota division, is commonly associated with vaginal infections. This yeast is dimorphic, producing characteristic pseudomycelia. Fungal diseases are called *mycoses.*

Fungi are classified by the type of sexual spores they produce. There are four main divisions:

1. *Zygomycota or Phycomycota.* These molds produce asexual *sporangiospores* in compact sacs called *sporangia.* The sporangia are borne on the tips of reproductive, aerial hyphae (called sporangiophores),

which extend aerially above the surface of the vegetative mycelium. The sacs burst, and each sporangiospore is capable of forming a new mold on an appropriate substrate. In sexual reproduction, the hyphae from one mycelium contacts and fuses with the hyphae from another mycelium and forms a *zygospore.* This *zygospore* can then yield asexual spores that can form new molds. The mold *Rhizopus* is an example of this class. (See Fig. 7.2a.)

2. *Ascomycota.* These molds produce asexual *conidiospores,* now known as *conidia.* Conidia are borne externally on aerial hyphae called conidiophores. In sexual reproduction, they form *ascospores* in a sac called an *ascus.* This division includes *Penicillium, Aspergillus, Blastomyces, Histoplasma,* and the bread yeast, *Saccharomyces.* (See Figs. 7.2b, c, d.)

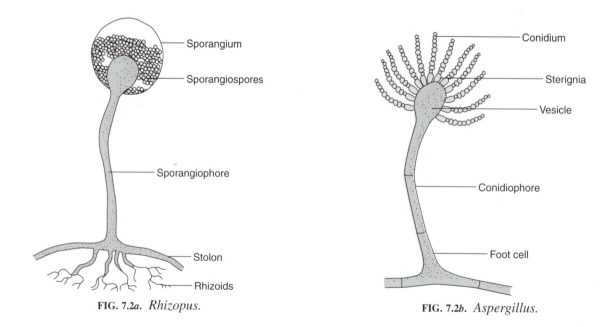

FIG. 7.2a. *Rhizopus.*

FIG. 7.2b. *Aspergillus.*

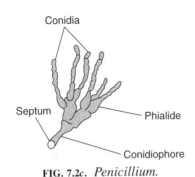

FIG. 7.2c. *Penicillium.*

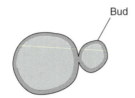

FIG. 7.2d. *Saccharomyces.*

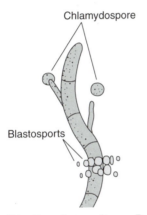

FIG. 7.2e. Candida *Pseudomycelia or Pseudohyphae.*

3. *Basidiomycota.* These molds produce sexual *basidiospores* on club-shaped *basidia.* Common mushrooms, toadstools, and puffballs are all members of this group.

4. *Deuteromycota.* These molds are referred to as imperfect fungi (*Fungi imperfecti*) because their sexual stage of reproduction is unknown. A number of these have been reclassified as Ascomycota. *Candida* is a member of this division. (See Fig. 7.2e.)

Materials List Per Table/Workstation

Sabouraud dextrose agar plates and prepared slides of *Rhizopus nigricans, Penicillium notatum, Aspergillus niger*

Prepared slide of *Rhizopus* zygospores

Broth culture of the yeast *Saccharomyces cerevisiae*

Prepared slide of the yeasts *Candida albicans* or *S. cerevisiae*

Samples of moldy food

Magnifying glass or stereo microscope

Dilute methylene blue solution

PROCEDURE

1. Examine Petri dish cultures on Sabouraud dextrose agar of the following molds: *Do not open these dishes!* Their spores are easily airborne and can cause infection, allergy, or laboratory contamination.

   a. *Rhizopus nigricans,* a black, bread mold

   b. *Penicillium notatum,* a blue-green, bread mold

   c. *Aspergillus niger,* a black, grain mold

   Note their mycelial masses, vegetative hyphae, and reproductive aerial hyphae.

2. Examine prepared slides of *Rhizopus* and the *Rhizopus* zygospore. Sketch the sporangia, sporangiophores, sporangiospores, and the zygospore; any rhizoids observed; and their nonseptate hyphae.

3. Examine prepared slides of *Penicillium* and *Aspergillus.* Sketch their septate hyphae, conidiophores, conidia, and any rhizoids.

4. Examine any examples of moldy food available. *Do not open!*

5. Place one or two drops of dilute methylene blue solution onto a glass slide. Aseptically place a loopful of yeast broth culture onto the stain. Do not heat fix. Place a cover slip on the slide. Examine both this live culture and any prepared yeast slides for the oval yeast cells. Note the presence of a nucleus in these cells and large storage vacuoles. Look for budding. Sketch what you observe.

Results

Draw your observations of the following Fungi:

Rhizopus sporangia, sporangiophores, sporangiospores, rhizoids, and zygospores

Penicillium septate hyphae, conidiophores, conidia, and rhizoids

Aspergillus septate hyphae, conidiophores, conidia, and rhizoids

Saccharomyces budding

NAME _____ DATE _____ SECTION _____

QUESTIONS

1. According to what features are the fungi classified?

2. How do fungi reproduce?

3. How do yeasts reproduce?

4. How do you differentiate between a yeast cell and a bacterial coccus?

MATCHING

a. dimorphic

b. no sexual reproduction

c. *Penicillium*

d. *Candida*

e. rhizoid

f. mushroom

g. *Rhizopus*

h. hyphae

_____ Zygomycota

_____ Ascomycota

_____ Basidiomycota

_____ Deuteromycota

_____ rootlike system

_____ exists in both mold and yeast forms

_____ pathogenic yeast

_____ threadlike filaments

(Answers may be used more than once.)

MULTIPLE CHOICE

1. Yeasts are:

   a. asexual spores   b. sexual spores   c. perfect   d. unicellular

2. Fungi possess all of the following except:

   a. nuclei   b. cell walls   c. chloroplasts   d. spores

3. Cross walls found in hyphae are called:

   a. cysts   b. mycelia   c. rhizoids   d. septa

4. Which of the following fungi would you most likely find in a salad (as a specifically added ingredient)?

   a. Zygomycota   b. Ascomycota   c. Basidiomycota   d. Deuteromycota

5. Which of the following fungi are commonly a source of antibiotics?

   a. Zygomycota   b. Ascomycota   c. Basidiomycota   d. Deuteromycota

6. Which of the following has no known sexual reproductive stage?

   a. Zygomycota   b. Ascomycota   c. Basidiomycota   d. Deuteromycota

7. The vegetative mycelia that mimics a "root" system is a/an:

   a. rhizoid   b. ascospore   c. ascophore   d. bud

WORKING DEFINITIONS AND TERMS

Aerial Growing into the air.

Bud Oval, asexual extension that pinches off of a yeast cell.

Dimorphic Able to grow in both the yeast and filamentous state.

Hypha A branching, threadlike, filament structure of fungi.

Imperfect Possessing no sexual stage.

Mycelium A mass of vegetative hyphae.

Mycology The study of fungi or the kingdom Myceteae.

Mycoses Diseases caused by fungi.

Rhizoids Rootlike structure for absorption of nutrients.

Saprophyte Living off decomposing matter.

Septate Divided by cross walls.

Zygospore Thick-walled sexual spore of Zygomycota.

Viruses—Visualization and Enumeration

Objectives

After completing this lab, you should be able to:

1. Describe what viruses are.

2. Differentiate between the various structural forms of viruses.

3. Properly prepare a serial dilution.

4. Identify a bacteriophage plaque.

5. Properly count bacteriophage plaques.

6. Determine the number of bacteriophages in a suspension.

VIRUSES

Viruses are among the smallest known pathogenic (disease-causing) agents. Viruses are not true cells; they are much smaller than cells and are often described as supra-molecules. Viruses generally are 0.2 μm or smaller in size. As a result, they are *filterable* (they pass through a membrane filter), and an electron microscope is needed to observe an individual viral particle, known as a *virion*. Viruses are traditionally described as "obligate intracellular parasites", meaning that they must use the enzymes and nucleic acids of the host cells they infect in order to reproduce. In order to grow viruses in a laboratory, you must first grow their hosts.

Viruses are host-specific and are first classified according to the organisms they infect. Viruses, which infect bacteria, are called *bacteriophages*. Viruses which infect humans are animal viruses or human viruses. Viruses are comprised of strands of nucleic acid, surrounded by a protein coat. The nucleic acid can be DNA or RNA, never both, and can be single stranded or double stranded. All these facts are used to further classify a virus. For example, *Herpesvirus*, the cause of the common cold sore, is a double-stranded DNA virus. The protein coat of viruses, called a *capsid*, can exist in either a *helical* shape or an *icosahedron* (a geometric shape of 20 triangular sides). The Herpesvirus is an icosahedral virus. Some animal viruses have an outer lipid envelope; some bacteriophages have tail structures attached to their capsids. (See Fig. 8.1.)

Procedure (Optional)

Examine various available electron micrographs of:

1. a helical human virus

2. an icosohedral human virus

3. a bacteriophage

BACTERIOPHAGE ENUMERATION

Animal viruses are typically grown in special *cell cultures,* that is, monolayers of specific host cells fed with liquid nutrients. Growing animal cells in such a culture is complicated and costly, and requires highly specialized equipment. Therefore, this lab will focus on the rapid growth of a bacteriophage, called T1, on an agar

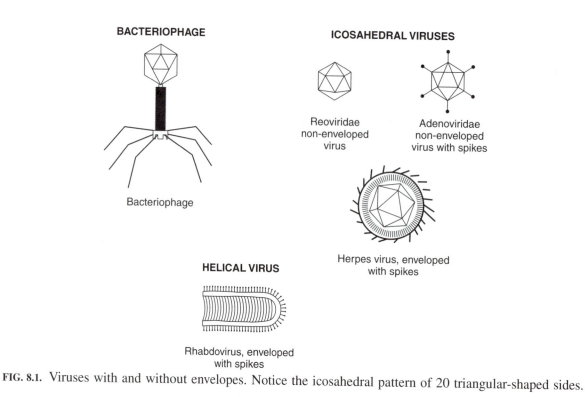

BACTERIOPHAGE

Bacteriophage

ICOSAHEDRAL VIRUSES

Reoviridae
non-enveloped
virus

Adenoviridae
non-enveloped
virus with spikes

Herpes virus, enveloped
with spikes

HELICAL VIRUS

Rhabdovirus, enveloped
with spikes

FIG. 8.1. Viruses with and without envelopes. Notice the icosahedral pattern of 20 triangular-shaped sides.

plate of its bacterial host: *E. coli* strain B. T1 is a *lytic phage;* that is, one initial phage will produce 200 to 300 new phage particles inside of a single host bacterial cell, causing the cell to burst or *lyse.* Each of these new phages is then capable of infecting 200 to 300 more host cells. You will seed soft agar with bacteria and successive dilutions of phage. The seeded soft agar mixture is then poured onto a base layer of nutrient agar and allowed to solidify. During incubation, bacteria grow throughout the soft agar overlay, yielding a cloudy, continuous surface or *lawn* of bacteria. Wherever a lytic phage has infected a bacterial cell, the continuous cycle of infection/lysis/reinfection/lysis produces a clear zone called a *plaque.* The soft agar prevents the unrestricted spread of the phage, ensuring that only bacteria adjacent to the initial infection are affected during incubation. (See Fig. 8.2.)

These plaques can then be easily counted, and the count can be multiplied by the reciprocal of the dilution

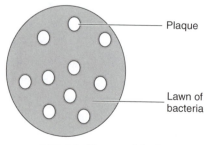

— Plaque

— Lawn of
bacteria

FIG. 8.2. Zones of lysis.

factor of the original phage inoculum. This readily yields an enumeration of the amount of phage originally present. This procedure is called a *plaque assay.* A similar plaquing assay is used clinically with human viruses whereby the tissue culture is overlaid with agar.

Materials List Per Student/Workstation

Six sterile saline (4.5 ml) tubes/group

Six nutrient agar plates/group

1 ml pipettes

T1 bacteriophage suspension (10^4/ml)

24-hour broth culture of *E. coli* B

Six tubes soft overlay agar/group (0.7% agar)

waterbath for melted agar, set at 45° C

PROCEDURE

1. Place six sterile saline (4.5 ml each) tubes in your test-tube rack.

2. Label one tube "control" and label the remaining five tubes consecutively from 10^{-1} through 10^{-5}.

3. Label six nutrient agar plates the same as the tubes.

4. Using a sterile 1 ml pipette, aseptically transfer 0.5 ml of the bacteriophage suspension provided to the saline tube labeled 10^{-1}.

 Do Not Pipette By Mouth. Always Use A Pipetting Bulb.

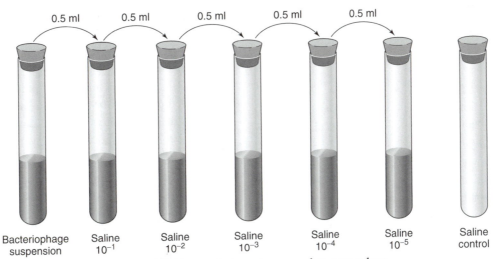

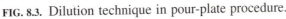

0.5 ml	0.5 ml	0.5 ml	0.5 ml	0.5 ml	

Bacteriophage suspension   Saline 10^{-1} Saline 10^{-2} Saline 10^{-3} Saline 10^{-4} Saline 10^{-5} Saline control

FIG. 8.3. Dilution technique in pour-plate procedure.

5. Mix the tube well by rolling it between the palms of your hands.

6. With another 1 ml pipette, transfer 0.5 ml from the 10^{-1} tube to the 10^{-2} tube. Mix the tube well as in step 5.

7. Using a fresh pipette for each transfer, transfer 0.5 ml of the suspension from the 10^{-2} tube to the 10^{-3} tube, and continue this diluting procedure consecutively for the remaining saline tubes. Don't forget to mix each tube well before and after diluting. You have now made a series of tenfold dilutions of the original phage suspension. (See Fig. 8.3.)

8. *(Note: You must work quickly here.)* Obtain six tubes of melted soft overlay agar from the waterbath.

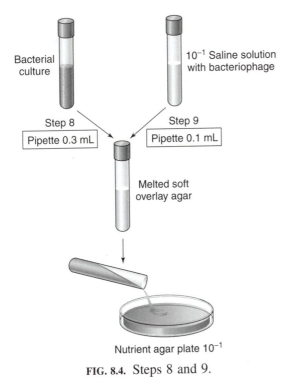

FIG. 8.4. Steps 8 and 9.

Pipette 0.3 ml of a broth culture of *E. coli* into each of the soft agar tubes. Mix each tube well by rolling between your palms. Label each tube with your initials and return them to the waterbath as soon as possible. Do not allow the agar to solidify.

9. *(Again work quickly.)* Remove one inoculated tube of soft agar from the waterbath. Wipe off all of the water from the surface of the tube. Using a 1 ml pipette, aseptically transfer 0.1 ml of the 10^{-1} saline phage dilution into the soft agar tube. Mix the agar tube by rolling it between your hands. (See Fig. 8.4.)

10. Immediately, aseptically pour the soft agar onto the surface of the nutrient agar plate correspondingly labeled 10^{-1}. Replace the lid and, without picking up the plate, rotate it gently in a 6- to 8-inch circle on the surface of the table to evenly distribute the agar. (See Fig. 8.5.)

11. Using a fresh 1 ml pipette each time and working quickly, repeat steps 9 and 10 for the remaining saline phage dilution tubes and for the saline control tube. For each dilution tube, use its correspondingly labeled nutrient agar plate.

FIG. 8.5. Rotation of melted agar.

12. Allow the soft agar to solidify.

13. Invert and incubate the plates at 35–37° C for 24 hours.

Results

1. After incubation, examine each plate and count the number of plaques on each plate that has clearly differentiated plaques.

2. Record your counts in Table 8.1. Plates where plaques have covered the entire plate and where plaques are not clearly discernible from each other (more than 300 plaques) should be recorded as TNTC (too numerous to count).

3. Note that the number of plaques recorded under the 10^{-3} column should be 10 times the number recorded under the 10^{-4} column; the number of plaques recorded under the 10^{-4} column should be 10 times the number of plaques recorded under the 10^{-5} column. The control should not show any plaques.

4. Calculate the number of lytic phages per milliliter that were in the original bacteriophage suspension using the following formula.

(*Note:* The number of plaques on each dilution plate is multiplied by 10 because only 0.1 ml of the saline phage dilution was transferred to soft agar in step 9, and plaques are always expressed in numbers per milliliter.)

$$\text{Plaque-forming units/ml} = \frac{\text{number of plaques on dilution plate} \times 10}{\text{dilution factor of the plate}}$$

TABLE 8.1

RESULTS OF BACTERIOPHAGE ASSAY					
Dilution of phage	10^{-1}	10^{-2}	10^{-3}	10^{-4}	10^{-5}
Number of plaques					
Calculations of plaque units/ml					

NAME _____ DATE _____ SECTION _____

QUESTIONS

1. What procedure is followed to make a tenfold dilution?

2. Do viruses possess enzymes?

3. How are viruses classified?

4. What is a bacteriophage?

5. Distinguish between a helical and an icosohedral virus.

6. How does a plaque develop?

7. Do viruses pass through bacteriological filters?

8. What do you expect to see on the control plate of this exercise? Why?

MATCHING

a. virion

_____ a virus that infects a bacterial cell

b. plaque

_____ an individual virus particle

c. phage

_____ the outer protein coat of a virus particle

d. capsid

_____ a clear zone of lysis on agar due to successive infection of host cells

e. icosohedral

_____ a geometrical shape of 20 triangular sides

f. plaque assay

g. capsule

MULTIPLE CHOICE QUESTIONS

1. Forty plaques were detected on a 10^{-3} dilution plate. The original concentration of the sample was _____ per ml.

   a. 400   b. 4000   c. 40,000   d. 400,000

2. Which statement is false about viruses?

   a. They are the smallest form of cells capable of causing infection.
   b. They must have living cells as hosts.
   c. They cannot be seen with conventional light microscopes.
   d. They are covered by protein.

3. The covering material of a viral particle is called a(n):

   a. icosohedron   b. capsule   c. capsid   d. phage

4. A 20-sided virus is termed:

   a. phage   b. convex   c. icosohedral   d. radial

5. The term *phage* in bacteriophage means:

   a. dissolve   b. puncture   c. destroy   d. eat

6. A viral suspension of 3800 per milliliter is plated out using the methods of this laboratory. Which of the following dilutions will yield accurate results?

   a. 10^{-1} b. 10^{-2} c. 10^{-3} d. 10^{-4}

WORKING DEFINITIONS AND TERMS

Bacteriophage A virus that infects a bacterial cell.

Capsid The outer protein coat of a virus particle.

Cell culture A single layer of animal host cells, nutrient-fed, used to maintain a virus.

Filterable Able to pass through a membrane filter that will prevent the transmission of bacteria.

Lytic phage A phage that causes the host cell to burst.

Obligate Requiring a specific condition.

Parasite Living in/on a host, at the expense of that host.

Pathogenic Disease causing.

Plaque A clear zone of lysis due to successive infection of host cells by a virus.

Tissue or cell culture A single layer of host cells, nutrient-fed, used to maintain a virus or increase viral numbers.

Virion An individual particle.

9 Parasitology

Objectives

After completing this lab, you should be able to:

1. Describe the relationship between a parasite and a host.
2. Differentiate between protozoan cysts and trophozoites.
3. Explain why the female *Anopheles* mosquito is considered a biological vector.
4. Explain why *Toxoplasma gondii* is dangerous to fetuses and people with damaged immune systems, but not as dangerous to people with normally functioning immunity.
5. Describe the method of transmission of *Giardia* and *Cryptosporidium*.

Any organism (or virus) that lives off another organism becomes a *parasite* if that other organism, called the *host,* suffers significant damage from this relationship. Parasitism is one of three forms of "living together," or *symbiosis. Mutualism,* in which both organisms benefit, and *commensalism,* in which one of the organisms benefits and the other is neither helped nor harmed, are the other two. Although bacteria, viruses, and fungi all have members that are quite capable of causing severe damage to their respective hosts, a more narrow view limits the scope of parasitology to protozoa, helminths (worms), and arthropods (insects and arachnids). This exercise will consider several protozoan parasites.

Most protozoan parasites associated with human diseases belong to the Class *Sporozoa.* Characteristics of this group include the alteration of sexual and asexual stages of development and a resistant sporelike stage of development called a *cyst* or an *oocyst.* The actively growing and motile forms of these microbes are called *trophozoites.* During this trophozoite stage, the protozoan reproduces, usually asexually, and invades the host tissue. This is also the stage that is most easily killed by body defenses, changes in the environment, or med-

ication. The cyst, or oocyst, can resist drying and harsh chemicals for long periods of time. *Cryptosporidium* cysts easily survive the levels of chlorination found in municipal water supplies.

Protozoan parasites are responsible for many diseases in tropical and subtropical areas of the world. Malaria is making a serious comeback as a major disease after years of control via insecticides and drugs. Now resistance to both of these control methods is making malaria a killer of over 2 million people per year.*

Closer to home, protozoans such as *Giardia, Cryptosporidium,* and *Toxoplasma* cause serious diseases among various populations in the United States. *Giardia lamblia* is frequently found in untreated water supplies such as rivers and streams. Hikers who pause for that refreshing cool drink may wind up with several days of cramps, diarrhea, nausea, and flatulence. Even more prevalent now is *Cryptosporidium parvum,* which has been able to invade the water supplies of major

---

*Malaria is typically thought of as a foreign disease, but it has often shown up in the United States. Most recently, in 1988, two boys contracted this disease while in a New York State Boy Scout camp.

cities. The city of Milwaukee was placed at risk several years ago when its entire water supply became contaminated.

A significant risk to fetuses is *Toxoplasma gondii*. Infection from this microbe can be traced to undercooked meat and cat litter boxes. Up to 50% of fetuses whose mothers were infected during the first trimester of pregnancy become infected themselves. (See Fig. 9.5 for the life cycle of this microbe.) The possible consequences of such prenatal infection include miscarriage, stillbirth, and numerous congenital defects, including mental retardation. The most common defect is symptoms of *retinitis*, which involves pain, light sensitivity, and blurred vision. The parasite can remain latent in the body for years before it becomes active. Depression of the host's immune system acts as a trigger for this activation; therefore, *Toxoplasma gondii* is often one of the diseases found in full-blown or frank AIDS patients.

Materials

Prepared slides of *Plasmodium vivax* in blood; *Toxoplasma gondii* oocysts, pseudocysts, sexual and asexual forms, trophozoite; *Giardia lamblia* trophozoite, sporozoite; *Cryptosporidium parvum* oocyst, sporozoite with merozoites

Malaria has a complex life cycle that includes the salivary glands of the *Anopheles* mosquito as well as the red blood cells of humans. The mosquito is considered an example of a *biological vector*. The *Plasmodium* parasite must spend part of its life cycle within the mosquito in order to become infective to humans. Therefore, if you can control the vector, you can control the disease. Although primarily a tropical disease, two species, *Plasmodium falciparum* and *Plasmodium vivax,* are found in the United States. *Plasmodium vivax,* the more virulent form, will be used as an example of the life cycle. (See Fig. 9.1.)

Inside the intestine of the female *Anopheles* mosquito, male and female forms of the parasite called *macrogametocytes* (female) and *microgametocytes* (male) combine to form a *zygote,* or fertilized egg. Unfortunately for some unsuspecting mammal, including humans, a blood meal is required for this process to take place. The zygote matures into a wormlike form, which then develops into an *oocyst*. Within the oocyst, hundreds of infective *sporozoites* develop. When the

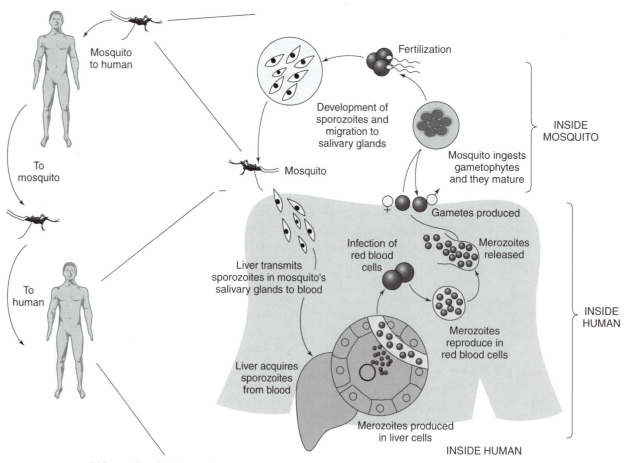

FIG. 9.1. Life cycle of *Plasmodium vivax*. Note the dependence on the female *Anopheles* mosquito.

MICROBIAL MORPHOLOGY, DIFFERENTIAL STAINS

oocyst lyses, the sporozoites spread throughout the mosquito, including its salivary glands and ducts.

When the mosquito feeds again, some saliva, acting as an anticoagulant, enters the feeding site of the mammal. The sporozoites are carried to the liver where they invade the host's liver cells, reproduce, leave, and then infect red blood cells. Once inside the red blood cell, the sporozoite develops into a ringlike *trophozoite* which now develops into thousands of infective *merozoites*. The red blood cell lyses, and the merozoites infect adjacent cells. During this process, some trophozoites develop into the male and female gametocytes, which are also released into the blood stream. The cycle continues if the infected individual is once again bitten by a female *Anopheles* mosquito.

Observe the slide of *P. vivax* under oil immersion. Find and identify the forms shown in Figs. 9.2, 9.3, and 9.4.

Toxoplasma gondii has a life cycle that invariably includes cats, which are its primary host. (See Fig. 9.5.) The microbe can be found in undercooked meat such as pork (25%) and lamb (10%). From this reservoir, it can infect humans directly, or it can first infect a household cat. Cats can also become infected by eating infected mice and rats.* Since humans are not part of this para-

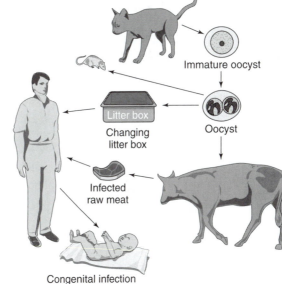

FIG. 9.5. Life cycle of *Toxoplasma gondii*.

FIG. 9.6. Small intestine of cat showing sexual and asexual forms.

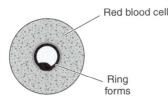

FIG. 9.2. Ringlike trophozoite.   FIG. 9.3. Merozoites.

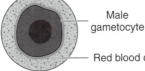

FIG. 9.4. Microgametocytes and macrogametocytes.

FIG. 9.7. Diagram of oocysts and sporozoite forms.

site's normal life cycle, we are considered *accidental hosts*.

When reproducing in the intestinal tract of the cat, some of the microbes differentiate into male and female gametes, called micro and macro gametocytes, the equivalent of eggs and sperm. (See Fig. 9.6.) When these gametes unite, a thick-walled *oocyst* is formed, which is then expelled in the feces by the millions. In the soil, each oocyst develops into two oocysts, each containing four *sporozoites*. (See Fig. 9.7.) Once the sporozoites form, the microbe is infectious. Since the oocysts are resistant to drying and remain viable for up to a year, they can waft into the air when cat litter is changed. They undergo further maturation and eventually become ingested or inhaled.

*Recent research suggests that this parasite enhances its chance of survival while in one of its intermediate hosts, the rat. When the rat is infected, usually by contacting the cyst form of *Toxoplasma* from soil or food, the brain becomes damaged. Such brain damage causes the rat to be less aware of its surroundings, including the scent of cats and the odor of its urine. This phenomenon allows the rat to be more likely eaten by cats, thus continuing the life cycle of the parasite.

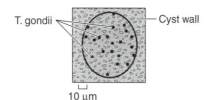

FIG. 9.8. Trophozoite or tachyzoites within pseudocyst in liver.

FIG. 9.9. Pseudocysts in brain.

Once the oocysts and sporozoites are ingested by other animals, including humans, they invade the cells of the intestines, and spread to the cells of the heart, brain, and muscle tissue. In these other animals, there is no sexual stage as in the cat. As host immunity is stimulated, large numbers of *tropozoites* (also called *tachyzoites*) become contained within protective coverings produced by the host called *pseudocysts*. (See Fig. 9.8.) As long as the immune system remains efficient, these pseudocysts remain intact, and further spread of the trophozoites is blocked. If a person with pseudocysts has a severe immunodeficiency such as AIDS or has a poorly developed immune system, such as a fetus, the pseudocysts either never develop or they break open, allowing the parasite to spread. (See Fig. 9.9.)

Giardia lamblia was first described by Leeuwenhoek over 300 years ago. (See Fig. 9.10 for life cycle.) Students are often startled at their first look at this parasite whose arrangement of paired nuclei in the *trophozoite* form gives the appearance of them looking back. (See Fig. 9.11.) The trophozoite parasitizes the upper portion of the small intestine where it holds on to the intestinal wall by way of an adhesive disk, much

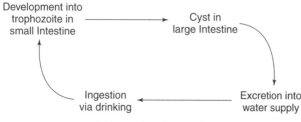

FIG. 9.10. Life cycle of *Giardia lamblia.*

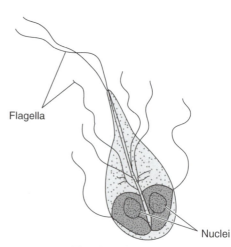

FIG. 9.11. Trophozoite of *G. lamblia.*

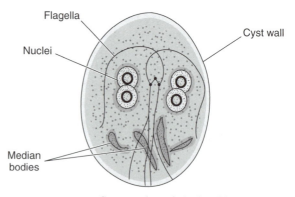

FIG. 9.12. Sporozoite of *G. lamblia.*

like a suction cup. If they break loose, they are carried toward the colon where many of them develop into the inactive *cyst* form. (See Fig. 9.12.) It is this cyst that causes disease when ingested. If the trophozoite form is swallowed, it will not survive the acidity of the stomach. (See Fig. 9.10.)

Cryptosporidium is also an intestinal parasite. Its method of transmission is similar to that of *Giardia* (fecal-oral), but it is even more resistant to control methods such as chlorination. Its life cycle is similar to that of *Toxoplasma,* with various animals acting as intermediate hosts. (See Fig. 9.13.) While only annoying to healthy people, severe, uncontrollable diarrhea and death can be the result in AIDS patients. Young children and other immunocompromised people are also at high risk for severe symptoms.

The human becomes infected by drinking water contaminated with the cryptosporidia oocysts or by eating food prepared with such water. The usual host for this parasite includes cattle, with up to an 80% infection rate. Poultry, sheep, even puppies and kittens, also

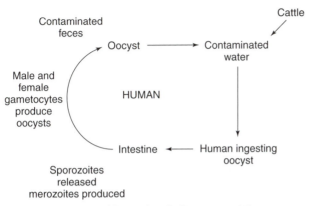

FIG. 9.13. Life cycle of *Cryptosporidia.*

9.14.) Some merozoites develop into male and female gametes. When these gametes combine, they then produce the resistant oocysts. The oocysts then leave via the feces and the cycle continues. (See Fig. 9.15.)

FIG. 9.14. Late stage of sporozoite development with eight banana-shaped merozoites.

show significant rates of infection. Once in the intestine, the oocyst releases sporozoites, which then invade the intestinal wall. The sporozoites divide into merozoites, which continues the invasive process. (See Fig.

FIG. 9.15. Oocyst of *Cryptosporidium.*

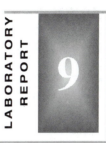

NAME _____ DATE _____ SECTION _____

QUESTIONS

1. What organisms other than protozoa would be considered parasites?

2. Why would proper uses of insecticides, proper water purification methods, and proper meat and poultry inspection and handling significantly reduce the numbers of parasitic infections or infestations?

3. Differentiate between the asexual stage of parasitic infections and the sexual stage.

4. Why would changing kitty litter be a possible danger to a pregnant woman?

5. Why are parasitic diseases a greater threat to the immunocompromised than to those with a normal immune system?

6. Differentiate between an intermediate host and a definitive host.

MATCHING

a. accidental host of *T. gondii*

b. zygote

c. cyst

d. biological vector

e. definitive host of *T. gondii*

f. pseudocyst

g. gametocyte

h. trophozoite

i. *Giardia lamblia*

j. *Cryptosporidium parvum*

_____ active form of protozoan parasite

_____ resistant form of protozoan parasite

_____ intermediate animal or host needed for a parasite to complete its life cycle

_____ fertilized egg

_____ male and/or female form of a parasite

_____ humans

_____ cats

_____ "eyelike" paired nuclei

MULTIPLE CHOICE

1. The sexual reproductive stage of *Plasmodium* takes place:
   a. in the human liver
   b. in red blood cells
   c. in a mosquito's intestine
   d. in blood plasma

2. The first site to be infected in a human by *Plasmodium vivax* is:
   a. the human liver
   b. red blood cells
   c. nervous tissue
   d. blood plasma

3. The merozoite form of *Plasmodium vivax* is found:
   a. in the human liver
   b. in red blood cells
   c. in a mosquito's intestine
   d. in blood plasma

4. The oocyst of *T. gondii* can be found:
   a. in cat litter boxes
   b. in the soil
   c. in the air
   d. in all of these

5. The definitive or final host of *T. gondii* is:
   a. cat   b. human   c. cow   d. rat

6. Which of the following can cause fetal damage?
   a. *Plasmodium*   b. *Toxoplasma*   c. *Giardia*   d. *Cryptosporidium*

7. Which of the following is found in contaminated water?
   a. *Plasmodium*   b. *Toxoplasma*   c. *Giardia*   d. all of these

8. Eight banana-shaped merozoites can be observed:
   a. in the macrogametocyte of *Plasmodium*
   b. in the sporozoite of *Cryptosporidium*
   c. in the trophozoite of *Giardia*
   d. in the pseudocyst of *Toxoplasma*

WORKING DEFINITIONS AND TERMS

Biological vector An animal, such as a mosquito which allows a parasite to spread, in which the parasite must spend part of its life cycle and where it is able to reproduce.

Host The organism in or on which a parasite lives, often causing harm or disease.

Intermediate host An animal in which the parasite goes through a developmental stage.

Macrogametocyte The female gametocyte of the sexual stage of protozoan reproduction.

Merozoite The motile, infective stage of sporozoan protozoa.

Microgametocyte The male gametocyte of the sexual stage of protozoan reproduction.

Oocyst The encysted form of a fertilized zygote or egg. The oocyst tends to be resistant to disinfection and releases large numbers of infectious sporozoites.

Parasite An organism that lives on or in another, derives nourishment, and often causes harm or disease.

Trophozoite The ameboid, asexual form of certain single-celled parasites.

PART III

MICROBIAL CONTROL AND BIOCHEMISTRY

The use of bacterial cells in the study of biochemistry has been well established for over a century. You may still have fond memories of the wonders of the Krebs cycle or oxidative phosphorylation* from other courses you may have taken. Much of the work done on these phenomena was originally done using bacteria. Prokaryotic cells are often used to study biochemical reactions because these cells are relatively easy to grow and their substrates, products, and enzyme reactions are easy to recognize.

In the field of microbiology, many of these biochemical reactions can determine whether a specific microbe resists the action of certain disinfectants and antibiotics. In the field of medical microbiology, many of these biochemical reactions are used to determine whether a certain bacterium is considered dangerous (virulent). If we can determine the exact biochemical reactions of a specific microbe, we can usually identify that microbe.

*These chemical reactions established how energy is derived from the breakdown of a glucose molecule and the transfer of much of that energy into the production of ATP.

Most of the laboratory exercises for the rest of the course will be concerned with aspects of bacterial biochemistry, with some dedicated to bacterial identification.

Although the concept of studying biochemistry may initially be rather intimidating, once certain principles are understood, it tends to be rather easy. Much research has been done to make these tests easy to perform and easy to "read." Reading a reaction means looking at the results and determining whether or not a certain reaction took place. Most of the biochemical reactions you will be studying can be read at a glance. The basis of the reading is as follows:

A specific substrate is placed in a growth medium. A microbe is inoculated and allowed to grow. If the substrate is utilized or changed, the medium will change color due to the presence of other indicator chemicals. These indicator chemicals (think of litmus, which changes color based on pH) are either part of the medium's formulation or added afterward. In most cases, the color change is very obvious. A color change usually indicates a positive reaction.

Microbial Sensitivity Testing

EXERCISE 10

Objectives

After completing this lab, you should be able to:

1. Determine the advantages and disadvantages of using ultraviolet light as a sterilizing agent.

2. Determine which species of bacteria was better able to withstand ultraviolet light, and formulate an answer as to why.

3. Determine how evidence of mutations can be detected when a microbe is subjected to ultraviolet light.

4. Determine whether some species of bacteria can resist high temperature while others cannot, and formulate an answer as to why.

5. Relate the principle behind the Kirby-Bauer method of antibiotic sensitivity.

6. Determine what constitutes a susceptible or sensitive reaction, an intermediate reaction, and a resistant reaction with the Kirby-Bauer method of determining antibiotic sensitivity.

10

EXERCISE

One of the clinical microbiologist's major responsibilities is to determine the best way to control microbes both on environmental surfaces and within the body. Infectious agents on environmental surfaces have the potential of finding a pathway (or portal of entry) into an unsuspecting patient or hospital worker; thus, these microbes must be controlled. Physical and chemical agents are employed for the sterilization and disinfection of microbes. Not all agents have the same effect (the rate of microbial death varies with the agent used and the type of microbe). The technique selected must be appropriate for the specific situation. Physical agents for the control of microbes include heat (moist and dry), pasteurization, freezing, radiation, and filtration. Chemical agents include a wide variety of antimicrobial substances and drugs, which must also be carefully selected for each situation. It is also the function of the clinical microbiologist to determine which antimicrobial drug,

for example, antibiotic, to recommend to the physician treating a patient with a bacterial infection. Aspects of controlling bacteria quickly and efficiently in the work area and within the body will be covered in this laboratory session.

PART I: PHYSICAL METHODS

ULTRAVIOLET LIGHT SENSITIVITY

Ultraviolet (UV) light damages cells in two ways: (1) it triggers mutations in DNA, resulting in thymine-thymine dimers, thus preventing successful reproduction, and (2) it causes direct protein damage, as can be seen in anyone suffering from sunburn. Since bacterial cells have only one chromosome, and even one mutation is often lethal, the large number of mutations caused by UV light often results in the death of practically all cells present. Since most cellular proteins are enzymatic, the few cells that may survive DNA damage will die as a result of enzyme damage. The combination of these two forms of cellular destruction often renders the surface of an object sterile when exposed to properly utilized ultraviolet radiation for sufficient time.

Most UV lamps used in the laboratory are not very powerful. Coupled with the factor that the distance between the light source and target organism will remain the same (these are constants), time will be the variable factor in determining how much UV exposure the microbes receive. Ultraviolet light cannot penetrate very well and is only effective on surfaces, so make sure the plates are uncovered while exposed to the light source. This form of radiation also converts atmospheric oxygen into irritating ozone and can damage the retina of the eye. Therefore, its use in the clinical area is somewhat limited.*

Sterilization by the Use of Ultraviolet Light

Materials List Per Table/Workstation

Broth cultures of *Serratia marcescens, Bacillus subtilis, Bacillus cereus*

Eight nutrient or T-Soy agar plates

One ultraviolet lamp

One pair of glasses (optional)

This procedure will be done per group or table. Before you start, turn on the ultraviolet lamp and allow it to warm up for several minutes. Make sure the bulb is working by reflecting the light against a paper towel. *Do not look directly at the light!* Then place it on the table so that the light is facing down.

PROCEDURE

1. Using aseptic technique, place a sterile swab in the *B. subtilis* broth. Withdraw the swab, and press and twist it on the inside of the tube above the level of broth to remove the excess liquid.

2. Draw the swab once across the agar on all seven plates as shown in Fig. 10.1. Discard the swab in disinfectant solution.

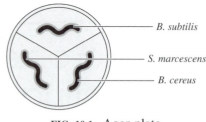

FIG. 10.1. Agar plate.

3. Repeat this procedure with the other two cultures. When finished, all plates will have a single line of bacteria spread across their surfaces. (See Fig. 10.1.)

Note: Make sure the three lines of inoculation are well separated from each other, especially in the center of the plate.

4. Label the plates 1 through 7 and place plate 1 aside. This will be the control. There should be good growth on all three lines of inoculation since this plate will not be exposed to any ultraviolet light. If there is no growth on any of the lines of inoculation of this plate, all results will be suspect.

5. Remove the cover of plate 2, place the bottom of the plate, agar side up, on the table, and place the ultraviolet light over it. Keep the light in place for the time stated in Table 10.1 or as directed by your instructor. (Because of differences in the power of the light, and distance between the light and the agar plate, these times may vary greatly.)

Note: You must remove the cover from the plate. Remember that ultraviolet light is low-energy radiation and has little penetrating power.

6. Repeat the procedure for the remaining plates using the exposure times in Table 10.1 or the times given to you by your instructor.

7. Incubate the plates upside down.

	Plate	Time	Observed Results
TABLE 10.1	1	Control 0 sec	
	2	20 sec	
	3	40 sec	
	4	60 sec	
	5	80 sec	
	6	5 min	
	7	30 min	

*UV light penetrates so poorly that regular window glass can prevent the penetration of over 80% of it. If this is the case, why is it so important to have sunglasses made of glass that is 100% UV resistant? Since the tinted glass prevents light from penetrating, the pupils respond by dilating. Without 100% UV protection, these dilated pupils would allow even more UV light to reach the retina.

The Penetrating Power of UV Light

As stated above, UV light has very little penetrating power. The cover of the agar plate, a glass slide, sunglasses, and even regular glasses will effectively interfere with the passage of UV light. This part of the exercise will demonstrate that phenomenon.

PROCEDURE

1. Aseptically remove a sample of broth from the *S. marcescens* broth tube using a sterile cotton swab.

2. Spread the bacteria across the entire surface of the agar plate with the swab. Make sure the swab completely covers the entire surface of the agar. This complete coverage of the agar plate is often called making a lawn. (See Fig. 10.2.)

3. Rotate the plate 60° to 90° and repeat step 2. (See Fig. 10.3.)

4. Rotate the plate again and repeat step 2 once more. Now the plate has been completely covered three times with the bacterial inoculum. (See Fig. 10.4.)

5. Go around the rim of the plate once or twice with the swab to make sure that there are no sections untouched by the inoculum. (See Fig. 10.5.)

6. Discard the used swab in a container of disinfectant.

7. Label the plate.

8. Place the agar plate under the UV light and you may now:

   a. partially remove the cover of the plate so the UV light must go through the cover to reach the plate as well as directly reaching the surface. OR

   b. place a slide so it is resting on the edge of the agar plate, once again allowing the light to penetrate the glass before it reaches the surface of the plate. OR

   c. carefully place a pair of eyeglasses between the plate and the light source. Be sure not to allow the glasses to touch the surface of the inoculated plate. (See Fig. 10.6.)

9. Expose the plate to UV light for 30 minutes, or as directed.

10. Incubate the plate upside down.

FIG. 10.2.

FIG. 10.4.

FIG. 10.3.

FIG. 10.5.

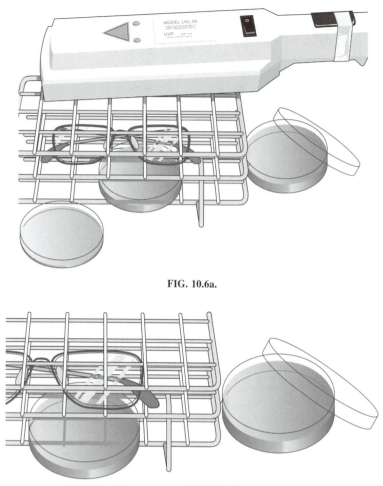

FIG. 10.6a.

FIG. 10.6b. Placement of UV light and agar plate

HEAT SENSITIVITY (May be done as an alternative to the UV light procedure)

As with other organisms, microbes have different tolerances to heat. Some are very sensitive to changes in temperature and are limited to a narrow range of temperature. They are thus *obligate* in this requirement. In other words, they must be kept at a fairly constant temperature for survival. For example, *Treponema pallidum,* the causative agent of syphilis, is normally found at human body temperature, which is 37° C. If the temperature drops to 20° C, which approximates room temperature, or if it rises to 40° C, which is consistent with a high fever, it quickly dies. Even the more temperature-tolerant or adaptive *(facultative)* ones used in this laboratory will have up to a 90% death rate if inoculated from room temperature into media taken directly from a refrigerator.

Thermal Death Time (TDT) is one of several methods used to explore the relationship between tempera-

ture, time, and the death rate of specific microbes. TDT is the time necessary to kill all vegetative cells in a pure broth culture at a predetermined temperature. As temperature goes up, the time necessary to kill microbes goes down. At 100° C, it takes only seconds to kill most vegetative cells. At 45° C, it may take an hour to kill many types of bacterial cells if they die at all.

Materials List Per Table/Workstation

Broth cultures of *Serratia marcescens, Bacillus subtilis, Micrococcus luteus*

10 ml nutrient broth tubes

One waterbath per class or per table

Six nutrient agar plates

PROCEDURE

1. Take six nutrient agar plates and divide them into eighths. Label two plates for inoculation with *Serratia marcescens,* two for inoculation with *Micrococcus luteus,* and the last two for inoculation

FIG. 10.7. Diagram of plate divided into eighths.

by *Bacillus subtilis*. Label each section of these plates starting at 0 and ending at 15. (See Fig. 10.7.)

2. Take the three tubes of inoculated broth plus a tube of sterile broth with the same volume of fluid and place them in a waterbath set to approximately 65° to 70° C. Place a thermometer in the tube of sterile broth to keep track of the temperature in the tubes. If a waterbath is available to each table, set the temperature differently for each group performing the exercise, for example, 60° C, 70° C, 80° C, 90° C, and 100° C. Record the exact temperature. It will take a few minutes for the temperature within the tubes to reach the temperature of the waterbath.

3. Before the tubes are allowed to warm up, aseptically take a sample of each microbe to be tested and inoculate the section of the nutrient agar plate labeled "0." You may do this procedure before the tubes are placed in the waterbath or immediately after, before the tubes have a chance to warm up. This is the control. Since the inoculum has not been subjected to any significant increase of temperature, this section of the plate should show growth. If there is no growth, all other results must be suspect. Make sure you do not overlap the inoculation into other sections of the plate.

4. Inoculate each section of the agar plate every minute and record the temperature within the control tube as it may fluctuate. Continue these inoculations every minute until all sections of the plates have a sample of the tested microbe in them. Make sure the broth is well mixed between each inoculation.

5. Incubate the plates upside down.

EFFECT OF COLD TEMPERATURE AND SLOW FREEZING (Optional)

Temperature-tolerant (facultative) microbes are not killed by a refrigerator temperature of 5° C. However, their metabolic reactions are significantly slowed down, as is their growth (reproductive rate). Slow freezing (−20° C) in the freezer compartment of a home refrigerator can cause large ice crystals to form, some of which may rupture cell walls and cell membranes of food as well as bacteria.

The cold temperature tolerance of some microbes will be tested by the following procedure.

Materials List Per Table/Workstation

Broth cultures of *Serratia marcescens, Bacillus subtilis*

10 ml sterile nutrient broth tubes

Calibrated loop

PROCEDURE

1. Take eight tubes of sterile nutrient broth. Label four tubes for inoculation with *S. marcescens*. In addition, label one tube "room temp," the second "refrig," the third "freezer," and the fourth "incubator." Repeat with four tubes for inoculation with *B. subtilis,* again labeling one for room temp, one for refrig, one for freezer, and one for incubator.

2. Gently mix the broth culture (by rolling the tube in an upright position between your palms) of *S. marescens* to achieve even distribution of bacteria throughout the broth and, using sterile technique, transfer one calibrated loopful (0.001 ml) of culture into each of the four tubes of sterile broth labeled *S. marcescens*. The object is to deliver equivalent numbers of bacteria into each tube. Again, using sterile technique, add one calibrated loopful of *B. subtilis* culture into each of the four tubes previously labeled.

3. Separate the tubes into four racks—one to be kept at room temperature, one in the refrigerator, one in a freezer compartment, and one to be incubated.

4. Each tube will be in the designated temperature condition overnight, and then all tubes will be refrigerated until the next laboratory period.

PART II: CHEMICAL METHODS

CHEMICAL SENSITIVITY

Microbial chemical sensitivity is the basis for the use of disinfectants and antiseptics on environmental and body surfaces. The thousands of different chemicals used in these products have all been tested to determine their safety, effectiveness, and usefulness under various conditions. One such test is a disk diffusion method performed on bacteria growing on an agar plate. Although the procedure is rather easy to perform, results are often difficult to interpret because the disinfectant is constantly in contact with the microbe rather than for a short exposure, and the microbe is on a growth medium rather than on a typical environmental surface, which has little nutritional material available.

The following procedure will indicate whether a chemical is effective against a certain microbe by readily demonstrating a *zone of inhibition* (or halolike area) surrounding the chemical on the agar plate. This zone of inhibition will have no bacterial growth.

Materials List Per Table/Workstation

Broth culture of one of the following: *Escherichia coli, Staphylococcus aureus, Pseudomonas aeruginosa*

Three nutrient agar plates

Sterile cotton swab

Forceps

12 assorted disinfectant and antiseptic solutions

Millimeter ruler

Paper Disks

PROCEDURE

Agar Plate Preparation. Choose one of the three broth cultures listed above and cover three nutrient agar plates with the culture following the procedure outlined for planting a "lawn" in the ultraviolet procedure above. (See Figs. 10.2 through 10.5.)

Placement of Disinfectants/Antiseptics. Once the three agar plates are prepared with the bacterial sample, aseptically place the disk-saturated chemicals on the agar surface as follows:

1. Subdivide the bottom of the three agar plates into four sections using a marking pen or pencil.

2. Dip the tip of the forceps into the alcohol solution and allow the alcohol to evaporate.

3. Remove a filter paper disk from the container and dip the disk *halfway* into one of the chemical solutions provided. (See Fig. 10.8.) Tap the disk on the side of the container to remove excess solution.

FIG. 10.8. Get sample of disinfectant or antiseptic by placing paper disk in the test solution.

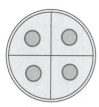

FIG. 10.9. Place 4 samples of the disinfectant or antiseptic test solutions on each agar plate.

4. Place the saturated disk in the center of one of the four sections on the plate and press down lightly with the forceps. Label the bottom of the plate to indicate which solution was used on each disk.

5. Repeat steps 2–4 with the other 11 solutions provided. If directed, use sterile water for one of your samples. This will act as a control.

6. When completed, each plate will have four disks of a different chemical solution diffusing into the growing bacteria. (See Fig. 10.9.) Afterward, place the plates in the incubation tray, upside down.

Note: Make sure everyone in your group knows which chemical is placed on each section of the plate.

7. After incubation fill out Table 10.2 and determine which chemicals were most effective based on the diameters of their zone of inhibitions.

TABLE 10.2

Chemical	Zone of Inhibition in mm

Chemotherapeutic Agent Testing: The Kirby-Bauer Plate

Before the advent of miniaturization and computerization of microbial techniques, one of the most common methods used to rapidly determine bacterial sensitivity and resistance to specific antimicrobial drugs was to use small paper disks, each saturated with a specific concentration of these different drugs. These disks are placed on an agar plate soon after the plate is evenly covered with the microbe being tested. This procedure is somewhat similar to the disinfectant testing method described earlier in the exercise. With the Kirby-Bauer procedure, however, standardization and mass production techniques allow hundreds of tests to be performed by a single laboratory worker. Standardization of the entire procedure is the key to ensuring accurate results.

Bacteria are standardized by placing them on the test medium in their early stages of growth. This ensures that all cells are equally susceptible to the antimicrobial agent. The concentration of these cells is controlled by comparing the cloudiness of the broth or saline solution that it is in with that of a standard chemical solution, which always displays uniform cloudiness. Originally, bacterial standardization was achieved by placing a sample from an isolated colony into a tube of broth and allowing it to grow for four to six hours until the cloudiness in the broth matched that of the control solution. Today, many laboratories modify this procedure by placing a larger sample of the inoculum into a tube of sterile saline or broth until the solution reaches the proper level of cloudiness. When placed on the agar plate, the cells go through the early stages of growth anyway and will be uniformly susceptible.

The antimicrobial disks are prestandardized at the pharmaceutical supply house. The amount of antimicrobial agent in each disk is exactly the same. For example, if penicillin G is one of the agents to be tested, each disk will contain exactly 10 milligrams of this antibiotic.

Finally, the culture medium used is also standardized in two ways. First, the type of medium used is of the same formula regardless of what laboratory performs the test. Of well over 100 different formulations available for laboratory use, only a few are used for this procedure. The second aspect of standardizing the growth medium is controlling the volume of material placed in each plate. The bacterial growth rate, the diffusion rate of the drug, and the amount of nutrients available to the growing bacteria are now all standardized.

Eight to 12 antimicrobial disks are placed on each plate of growth medium shortly after the inoculation of the bacteria. After incubation, the 8–12 disks placed on each Kirby-Bauer plate can be "read" or observed in

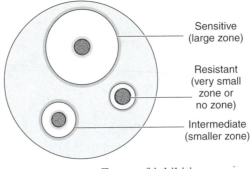

FIG. 10.10. Zones of inhibition.

seconds by an experienced microbiologist. The reading is based on the size of the zone of inhibition surrounding each disk. These zones are measured in millimeters (mm), and a difference in size of only 2–3 mm can mean the difference between describing an organism as being *susceptible* or *sensitive* to the drug, or being *resistant*, which indicates that the drug would be ineffective. The zone of inhibition that falls between that of susceptible and resistant is termed *intermediate*. (See Fig. 10.10.)

See Table 10.3 for examples of zone diameter and their interpretation.

THE KIRBY-BAUER TECHNIQUE

Materials List Per Table/Workstation

Agar plate with *Staphylococcus aureus, Escherichia coli, Pseudomonas aeruginosa*
Antibiotic dispensers
Three brain-heart infusion broth tubes
or
Three sterile saline tubes
Three Mueller-Hinton agar plates
Millimeter ruler

PROCEDURE

1a. Inoculate three broth tubes heavily with the three assigned bacteria. Label and place in the designated test-tube rack for incubation. Allow to incubate for approximately two hours. OR

1b. Use sterile cotton swabs to take samples of bacteria from each of the three assigned cultures. Place each sample in a tube of sterile saline solution and mix until slightly cloudy. These three swabs can be used for step 2.

2. Take a sample of broth (1a) or saline solution (1b) using a sterile swab, twist and press the swab on the inside of the tube above the liquid level to remove excess fluid, and remove the swab. Flame the tube and replace the cap.

Agent	*Disk Symbol*	*Potency*	*Resistant mm or Less*	*Intermediate*	*Sensitive mm or More*
Ampicillin	AM–10	10 μg			
For enteros (*E. coli*)			≤ 13	14–16	≥ 17
For staphylococci			≤ 28	—	≥ 29
Bacitracin	B–10	10 u	≤ 8	9–12	≥ 13
Carbenicillin	CB–100	100 μg			
For *P. aeruginosa*			≤ 13	14–16	≥ 17
For *E. coli*			≤ 19	20–22	≥ 23
Clindamycin	CC–2	2 μg			
For most organisms			≤ 14	15–20	≥ 21
Erythromycin	E 15	15 μg	≤ 15	16–20	≥ 21
For most organisms					
Kanamycin	K–30	30 μg	≤ 13	14–17	≥ 18
Methicillin	DP–5	5 μg			
For staphylococci			≤ 9	10–13	≥ 14
Oxacillin	OX–1	1 μg			
For staphylococci			≤ 10	11–12	≥ 13
Penicillin	P 10	10 u			
For staphylococci			≤ 28	—	≥ 29
Polymyxin B	PB 3000	300 u	≤ 8	9–12	≥ 12
Rifampin	RA 5	5 μg			
For most organisms			≤ 16	17–19	≥ 20
Streptomycin	S 10	10 μg			
For most organisms			≤ 11	12–14	≥ 15
Tetracycline	Te 30	30 μg			
For most organisms			≤ 14	15–18	≥ 19
Vancomycin	Va 30	30 μg			
For most gram positives			≤ 9	10–11	≥ 12

TABLE 10.3

3. Hold the Mueller-Hinton plate in your hand and cover the entire surface of the plate with your swab. This procedure can be done the same way as making the "lawn" with the chemical sensitivity testing. Another popular method of producing a lawn on larger plates is as follows:

  a. Smear the swab across the middle of the plate and then continue the smear in a zig-zag pattern across the plate away from you. Make sure you overlap previous areas and touch the sides of the plate as you go. (See Fig. 10.11.)

  b. Rotate the plate 90° and repeat the same procedure. When this step is completed, one-fourth of the plate will be covered twice and one-half of the plate once. (See Fig. 10.12.)

  c. Rotate the plate 90° again and repeat the procedure. Now one-half of the plate will be covered twice and one-fourth of the plate once. (See Fig. 10.13.)

  d. Repeat the procedure one last time. Now the entire plate has been covered twice.

  e. Draw the swab twice around the inner edge of the plate where the plate touches the agar, ensuring that every square millimeter of the agar is covered with a thin, even layer of bacteria. (See Fig. 10.14.)

Note: The entire surface of the plate must be covered to ensure accurate results.

4. Allow the broth or saline solution to absorb into the agar (one to two minutes) and place the antimicrobial disks evenly over the surface of the plate with a dispenser or forceps.

5. If a dispenser is used to place the disks on the plate, make sure you press each disk with a sterile loop or forceps to ensure proper or effective contact with the agar. If the dispenser happens to insert the disk sideways into the agar, flame a pair of forceps, allow it to cool, then remove the disk and place it on the agar properly.

6. When completed, label each plate and place them in the incubation tray, upside down.

Inventory

At the end of this exercise, each group will have:

Eight ultraviolet light agar plates and/or
Four Thermal Death Time agar plates
Eight nutrient broth tubes for cold sensitivity testing
Three disinfectant testing plates

And each person will have:

Three Kirby-Bauer antibiotic sensitivity plates

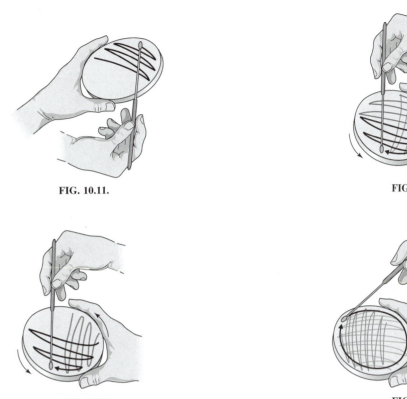

FIG. 10.11.

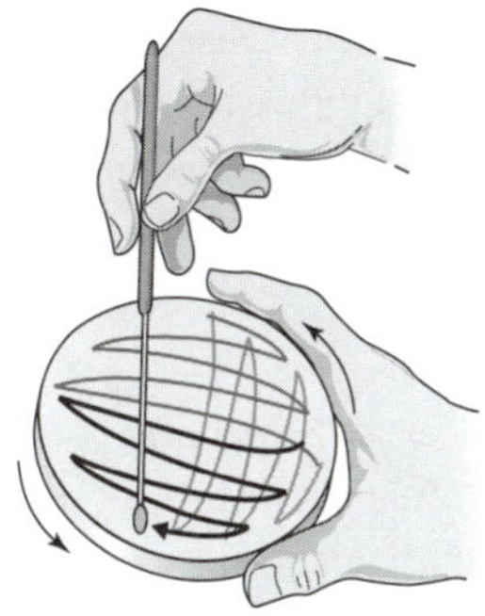

FIG. 10.13.

FIG. 10.12.

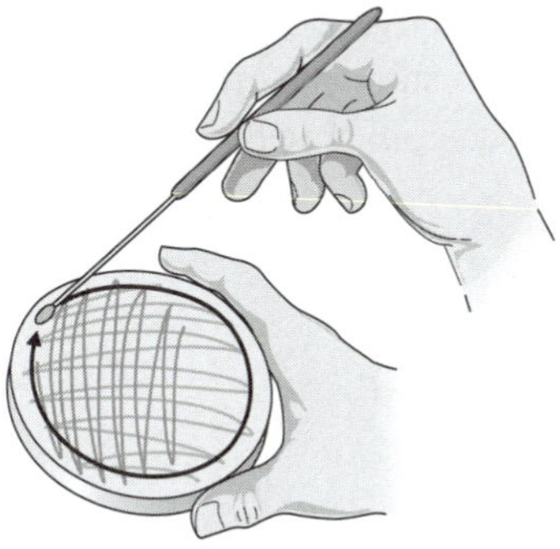

FIG. 10.14.

Results

Ultraviolet Light. Observe the ultraviolet light plates. Determine which of the inoculated microbes survived the longest. Based on previous laboratory exercises and lecture material, formulate an explanation. (*Hint:* Review previous labs on differential staining and/or lecture material regarding the characteristics of the genus *Bacillus*.) Were any color changes seen in any of the growth? What would cause any color change seen? (See Table 10.4)

TABLE 10.4	RESULTS OF KIRBY-BAUER ANTIMICROBIAL SENSITIVITY TEST					
Name of Antimicrobial	E. coli		S. aureus		P. aeruginosa	
	Zone of Inhibition in mm	Interpretation S = sensitive I = intermediate R = resistant	Zone of Inhibition in mm	Interpretation S = sensitive I = intermediate R = resistant	Zone of Inhibition in mm	Interpretation S = sensitive I = intermediate R = resistant

Thermal Death Time. Observe the Thermal Death Time plates. Was the temperature high enough to kill all the bacterial samples? Was one species of bacteria able to tolerate the heat better than the other two? (*Hint:* Review the UV light part of the exercise for the reason why.)

Effects of Cold and Freezing. After gently mixing the tubes, examine them for degrees of cloudiness, which increases with bacterial growth. Rank the four temperature conditions for each microbe, so as to establish the order of greatest amount of growth to least.

Chemical Sensitivity. Observe the nutrient agar plates prepared to show the effects of disinfectants and antiseptics on bacterial growth. Determine which chemical(s) were effective against the microbe by determining the relative sizes of the zones of inhibition on the plates. The size of the zone of inhibition is related to the diffusion rate of the chemical placed on the plate as it is in the Kirby-Bauer test.

Kirby-Bauer Test. Observe the Mueller-Hinton agar plates on which the Kirby-Bauer antimicrobial sensitivities were tested. Measure the zones of inhibition as directed. Determine which of the three microbes tested showed the greatest sensitivity to the drugs and which one showed the greatest resistance. You may see a colony within the zone of inhibition of a sensitive organism. Such a colony is the result of a strain of the test organism that has developed antibiotic resistance.

NAME _____ DATE _____ SECTION _____

QUESTIONS

1. Why should you avoid looking directly into the ultraviolet light?

2. Why is the method of testing chemical sensitivity to disinfectants you performed considered somewhat inaccurate?

3. Why did *Bacillus subtilis* show growth from all samples of the Thermal Death Time part of this exercise while the others soon showed no growth at all?

4. Why does milk that is pasteurized and then refrigerated have a limited shelf life and eventually "spoil"?

5. In the procedure used to test bacterial growth against various temperatures (incubator, room, refrigerator, freezer), why should efforts be made to inoculate each tube with the same number of bacteria?

6. Why does the Kirby-Bauer procedure require that the concentration of the bacteria be the same, the stage of growth constant, the growth medium the same, and the concentration or amount of drug in each disk constant?

7. Why are *Escherichia coli, Pseudomonas aeruginosa,* and *Staphylococcus aureus* used as standards in the Kirby-Bauer method?

8. A pure culture was inoculated onto a Mueller-Hinton agar plate. The Kirby-Bauer procedure was performed. One of the drugs tested showed a large zone of inhibition but also had small colonies growing within this zone. Further testing showed that these colonies were not the results of contamination. Why would these colonies be present within this zone of inhibition?

MATCHING

a. lawn

b. large zone of inhibition

c. small zone of inhibition

d. Thermal Death Time

e. Thermal Death Point

f. ozone

g. antiseptic

h. sanitizer

i. disinfectant

j. antibiotic

k. synthetic

l. semisynthetic

_____ a chemical that destroys most or all pathogens on an inanimate object

_____ solid growth of bacteria across the surface of a plate

_____ indicates that an antimicrobial drug would be effective against a specific microbe

_____ byproduct of UV light use

_____ amount of time it takes to kill 100% of a bacterial broth culture at a specific temperature

_____ indicates that an antimicrobial drug would not be effective against a specific microbe

_____ a substance naturally produced by one microbe that kills or inhibits another

MULTIPLE CHOICE

1. A disadvantage of using UV light to control microbes is:

   a. it produces irritable ozone
   b. it has poor penetrating power
   c. it can damage the retina of the eye
   d. all of these

2. UV light is able to damage bacterial cells by:

   a. preventing mitosis   b. damaging DNA   c. dissolving cell membranes   d. coagulating cytoplasm

3. *Bacillus subtilis* and *Bacillus cereus* most likely showed greater resistance to UV light than *Serratia marcescens*. This is due to:

   a. *Bacillus spp.* are Gram positive, while *Serratia marcescens* is Gram negative
   b. the *Bacillus spp.* are spore formers
   c. the pigment of *Serratia marcescens* absorbs more radiation than other cells
   d. *Bacillus spp.* have enzymes that protect them from ozone

4. The broth culture of *Bacillus subtilis* showed growth even after extended heating in a waterbath, while *Serratia marcescens* did not. This is due to:

   a. *Bacillus subtilis* are Gram positive, while *Serratia marcescens* is Gram negative
   b. the *Bacillus subtilis* are spore formers
   c. the pigment of *Serratia marcescens* absorbs more radiation than other cells
   d. *Bacillus subtilis* have enzymes that protect them from ozone

5. In the Kirby-Bauer test, which of the following must be consistent?

   a. concentration of bacteria placed on the plate
   b. type of medium in the plate
   c. concentration of antimicrobial drug in the disk
   d. all of these

6. A relatively large zone of inhibition surrounding an antimicrobial disk on a Kirby-Bauer test plate would most likely be interpreted as:

   a. sensitive reaction   b. intermediate reaction   c. resistant reaction

7. A factor in the zone of inhibition size on the Kirby-Bauer plate is:

   a. amount of medium placed on the plate
   b. rate of diffusion of the drug used
   c. stage of growth of the microbe placed on the plate
   d. all of these

WORKING DEFINITIONS AND TERMS

Antibiotic A substance naturally produced and released by one microbe that kills or inhibits other microbes. (Many antibiotics are prepared synthetically today.)

Antiseptic A chemical that can safely be used on the skin and mucus membrane which is able to destroy or inhibit most pathogens.

Bacteriocidal Refers to an agent that kills bacteria.

Bacteriostatic Refers to an agent that inhibits growth of bacteria.

Chemotherapeutic agent Any chemical that inhibits microbial growth within the body.

Disinfectant A chemical that destroys most or all pathogens on inanimate objects.

Intermediate reaction Zone of inhibition in the Kirby-Bauer procedure which indicates that the drug is only moderately effective against a specific infectious agent.

Mutation A permanent change in the sequence of nucleotides in a DNA molecule. Most mutations in prokaryotic cells are lethal to that cell.

Resistant Zone of inhibition reaction in a Kirby-Bauer procedure which determines that an antimicrobial drug is ineffective against a specific infectious agent.

Semisynthetic An antibiotic whose chemical components have been modified artificially.

Sensitive/susceptible Zone of inhibition reaction in a Kirby-Bauer procedure which determines that an antimicrobial drug is effective against a specific infectious agent.

Synthetic An antimicrobial drug produced completely under artificial conditions.

Thermal Death Time The time it takes to kill a pure broth culture of bacteria at a specified temperature.

Ultraviolet light A highly germicidal light with a wavelength of between 200 and 340 nanometers capable of damaging DNA and protein molecules.

EXERCISE 11

Bacterial Biochemistry

Objectives

After completing this lab, you should be able to:

1. Explain the concept of one enzyme, one substrate, one reaction.

2. Explain why certain microbes thrive in environments that are detrimental to other life forms.

3. Determine how to recognize whether a microbe possesses the enzyme to:

a. Ferment glucose and/or lactose and to produce acids and/or gas as a byproduct of the reaction.

b. Oxidize glucose.

c. Utilize citrate as a source of high-energy carbon.

d. Metabolize the amino-acid tryptophan into indole.

e. Remove the carboxyl group from an amino acid.

f. Reduce nitrates into nitrites and nitrogen gas.

g. Catabolize urea into ammonia by hydrolysis.

Bacteria can be found in virtually every environmental condition on Earth. They can be found thriving in sub-freezing Antarctic oceans as well as in the nearly boiling hot springs in Yellowstone National Park. Some grow in highly acidic pickle juice, and others are found in the alkaline springs of Death Valley. Some can even extract nutrients and growth factors from substances that would poison most other organisms (such as the oil spills infamous for destroying so many forms of wildlife).

Perhaps the major factor contributing to this tremendous range of survival is the vast array of enzymes available to them as a group. No one bacterium can survive all these conditions and use all nutrients available, but the variety of different bacterial cells ensures that no matter what the conditions, there is probably a bacterium with the necessary enzymes to ensure survival in that environment. This exercise will demonstrate that not all bacteria possess the same enzymes and that there are a wide variety of enzymes among bacteria. It will also introduce the concept that bacteria can be classified and identified based on the enzymes they possess.

CARBOHYDRATE METABOLIZING ENZYMES

Carbohydrates are used as high-energy sources by heterotrophic organisms, including most bacteria-colonizing humans. A single species of bacteria rarely possesses the enzymes that will catabolize or break down the over 40 different simple carbohydrates available as sources of energy. Enzymes work on very specific substances known as *substrates*. An enzyme reacts with one substrate only, such as glucose and not with any other, for example, lactose.* Therefore, if a bacterium is capable of metabolizing 27 different sug-

---

*This lack of an enzyme to properly metabolize lactose is commonly found in humans. Such individuals are "lactose intolerant." The sugar is a component part of dairy products such as milk and ice cream. Ingestion of such products allows the sugar to reach the lower intestine virtually unchanged. Unfortunately for such individuals, the bacteria that reside there have such an enzyme. Because of their ability to metabolize this sugar, significant amounts of gas are released as a byproduct, resulting in the production of significant amounts of flatulence by their host.

ars, it must be able to produce an enzyme that reacts with each one. The study of fermentation (an anaerobic reaction) and/or gas formation in glucose and lactose media will be part of this exercise. In addition, these different reactions allow us some insight into bacterial genetics. Enzymes are produced through protein synthesis. Since protein synthesis starts with a specific gene that is then transcribed and translated into an enzyme, identification of bacteria by the biochemistry of enzymatic reactions reflects the genetic makeup of each bacterium tested.

Metabolism may also be *aerobic* (oxidation) as well as *anaerobic* (fermentation). Therefore, the presence of oxygen can affect the results of the enzyme-substrate reaction. Oxidation-fermentation reactions involving glucose will be performed as part of this exercise.

Citrate is a carbohydrate substance, which some bacterial cells use as a source of high-energy electrons. Citrate is one of the molecules in the Krebs cycle or Citric Acid cycle, which may be familiar to you. It is also possible to determine whether a specific microbe contains the enzyme to use citrate as its only source of carbon for the generation of energy.

Materials List Per Table/Workstation

Agar cultures*: *Escherichia coli, Pseudomonas aeruginosa, Bacillus subtilis, Serratia marcescens*

Four tubes of Phenol Red Glucose broth

Four tubes of Phenol Red Lactose broth

Six tubes of Oxidation-Fermentation Basal medium

Four tubes of Simmons Citrate slants

Sterile Mineral Oil

Note: You will be using an inoculating needle for the oxidation-fermentation (O.F.) basal medium inoculation procedure. A needle *may* be used for most inoculating procedures, including slants, broths, and streak plates, but a needle *must* be used for procedures where the inoculum is to be placed below the surface of solid growth media.

PROCEDURE

1. *Carbohydrate fermentation.* Pick up and label four tubes of phenol red glucose and phenol red lactose

*Note to instructor: If this exercise is performed following Exercise 10, you may wish to use the control tubes of the UV light procedure and the Kirby-Bauer plates as the source of these cultures.

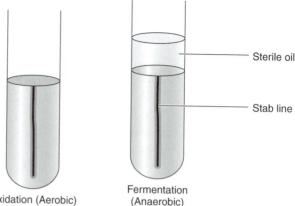

Oxidation (Aerobic)

Fermentation (Anaerobic)

Sterile oil

Stab line

FIG. 11.1. O.F. Basal test for oxidation.

FIG. 11.2. O.F. Basal test for fermentation.

broth. Phenol red is a pH indicator that turns yellow in an acid solution. Make sure you label the tubes of sugar as you pick them up; otherwise they will be indistinguishable from each other. Notice the small upside-down test tube in each of the larger tubes. These smaller tubes are *Durham tubes* and are used to trap gas. Using a loop, aseptically inoculate a sample of each of the assigned bacteria into each of the phenol red broth tubes. Make sure the tubes are labeled so that you will know which microbe has been inoculated in each tube.

2. *Oxidation-fermentation of carbohydrates.* Pick up six tubes of O.F. basal medium. Inoculate the tubes using an *inoculating needle* as directed. Inoculate two tubes with *E. coli,* two more with *Pseudomonas aeruginosa,* and the last two with *Bacillus subtilis.*

Once completed, aseptically place a few milliliters of sterile mineral oil in one set of O.F. basal medium. When completed, there will be one set of tubes of the three bacteria with mineral oil and another set without it. The set of tubes with the mineral oil will test for *fermentation* as the oil prevents the penetration of oxygen into the medium. The set of tubes without the oil will test for *aerobic oxidation* as air, and thus oxygen can reach the (surface of the) medium. (See Figs. 11.1 and 11.2.)

3. *Citrate utilization.* Pick up four tubes of Simmons citrate medium and inoculate each tube with each of the four assigned bacteria. Note the color of the slant. If citrate is utilized as the only source of carbon for high-energy production, a color indicator in the slant will cause a color change once growth occurs.

Results

"Read" the reactions in the tubes after incubation.

FIG. 11.3. Alkali (red). **FIG. 11.4.** Acid (yellow). **FIG. 11.5.** Acid + Gas (yellow + bubble in Durham tube).

Carbohydrate Fermentation (Phenol Red Broth) Tubes

Reactions

Alkali = red/orange tube (sugar was not fermented) + turbidity. (See Fig. 11.3.)

Acid = yellow tube (sugar was fermented) + turbidity. (See Fig. 11.4.)

Acid + Gas = yellow tube + gas present in the Durham tube. (See Fig. 11.5.)

O.F. Basal Medium. Arrange the tubes in pairs according to bacteria. One tube for each bacterium should have oil in it (test for fermentation), and the other should have no oil (test for oxidation). A color change from green to yellow indicates a pH change and thus utilization of the sugar.

No reaction = no color change (both tubes remain green)

Oxidation = tube without the oil will turn yellow (usually at the surface)

Fermentation = both tubes turn yellow

Note: Fermentation usually produces much more acid than oxidation. Even in the tube without oil, any subsurface growth would be anaerobic. Therefore, both tubes show the color change.

Simmons Citrate Medium. Observe the tubes and note the color change from green to blue, indicating citrate utilization. (Fill in Table 11.1)

AMINO ACID AND NITROGEN METABOLISM

Amino acids and other nitrogen-bearing compounds also form an important set of metabolites for microbes. As with the eukaryotic organisms, amino acids form the parts of cellular protein components such as enzymes. Certain microbes contain enzymes that can catabolize amino acids and use them as a source of energy.

Other bacterial types utilize nitrogen compounds completely differently than most eukaryotes do. In this way, nitrogen is constantly recycled from the gaseous form to the ionic form and back again. Bacterial enzymes are responsible for much of this recycling. Other

TABLE 11.1	RESULTS OF CARBOHYDRATE FERMENTATION					
	Growth Medium		*E. coli*	*P. aerug.*	*B. sub.*	*S. marc.*
	Phenol Red with Glucose	Acid				
		Gas				
	Phenol Red with Lactose	Acid				
		Gas				
	O.F. Basal with Glucose	Ox				
		Ferm				
	Simmons Citrate					

enzymes are capable of taking the nitrogenous waste product, urea, and catabolizing it into ammonia in order to extract energy. Such microbes are responsible for the ammonia odor of urine when it is allowed to stand at room temperature for long periods of time.

Materials List Per Table/Workstation

Agar cultures: *Escherichia coli, Pseudomonas aeruginosa, Bacillus subtilis, Proteus mirabilis*

Four tubes of Decarboxylase broth with Lysine, and/or

Four tubes of Decarboxylase broth with Ornithine, and/or

Four tubes of Decarboxylase broth with Arginine

Four tubes of Trypic Soy broth or Trypticase broth for indole production

Four tubes of Tryptic Nitrate medium for nitrate reduction

Two urea agar plates or slants or broths

Tubes of sterile mineral oil

Aluminum foil

PROCEDURE

1. *Decarboxylase.* Pick up and label four tubes of decarboxylase broth with the amino acid *lysine* and/or *ornithine* and/or *arginine* (total of four tubes of each type = 12 tubes). Make sure you label the tubes when you pick them up because they are indistinguishable from each other. Inoculate the tubes with the assigned bacteria and add a few milliliters of sterile mineral oil to each tube.

   The purple color of the broth is due to a pH color indicator. It is purple in an alkali environment and light yellow in an acidic one. *Decarboxylase* is an enzyme that will anaerobically (thus the oil) remove the *carboxyl group* (also called the carboxylic acid group) from the specific amino acid, for example, lysine, ornithine, or arginine.* In the case of arginine, this enzyme is often called *dihydrolase.* By removing the *acid* from an *amino acid,* the pH of the broth solution will rise or become more alkaline. The medium also contains glucose, which is rapidly fer-

---

*You may have indirect knowledge of these types of reactions if you have forgotten to refrigerate meat or any other food high in protein. The unpleasant odor emanating from this food is at least partially related to this breakdown of some of these amino acids. Cadaverine, as in cadaver, is the byproduct of the catabolism of lysine. Putrecine, as in putrid, is the byproduct of the destruction of arginine.

mented, causing the pH to initially drop. This rapid drop in pH will cause a color change from purple to yellow in approximately 12 hours. (If you wish to confirm this reaction, you may volunteer to return to the laboratory in 12 hours and note that all the tubes have turned yellow.) Decarboxylase works rather slowly and takes a full 24 hours of incubation to raise the pH. After 24 hours, any yellow tube indicates a negative reaction, and any tube "retaining" any shade of purple is considered positive. Remember, the enzyme decarboxylase works anaerobically, so make sure oil is added to these tubes.

2. *Indole production.* Pick up and label four tubes of tryptic soy broth or tryptone broth and inoculate them with the assigned bacteria. *Indole* is a byproduct of bacterial amino-acid metabolism. Certain microbes produce an enzyme that catabolizes the amino-acid *tryptophan,* which is found in tryptic soy broth and tryptone broth. When tryptophan is catabolized, indole is produced as a byproduct. Indole forms a red ring when mixed with *Kovac's reagent.* After growth occurs, carefully add several drops of Kovac's reagent to each tube.

3. *Nitrate reduction.* Pick up four tubes of tryptic-nitrate medium. Inoculate them with the assigned bacteria. This set of tubes will be used to test bacterial enzymes that reduce nitrate (NO_3) to nitrite (NO_2) and to further reduce nitrite to other compounds, usually, nitrogen gas (N_2). The reagents are sulfanilic acid (labeled Solution A) and N,N^1-dimethyl-alpha-naphthylamine (labeled Solution B). If these two chemicals are added to the tubes once growth occurs, the development of a red color will indicate the presence of nitrite.

Note: These chemicals are also somewhat toxic. If any are spilled on your hands, wash them off immediately.

No color change indicates that no reduction at all occurred and there is still nitrate in the tube, **or** the nitrate was reduced to nitrogen gas. *Powdered zinc* is added to the tubes displaying no color change. Zinc is a metal, and metals are *reducing agents.* If such an agent is added to tubes containing nitrate, the nitrate will soon be reduced to nitrite. Since both Solutions A and B are already in the tube, the medium in the tubes that still contain nitrate will soon turn red. The presence of a red color at this point indicates a *negative reaction,* for it was the zinc and not the bacterial enzymes that caused the reduction.

If there is no color change with the addition of zinc, it means there was no nitrate to reduce. Since no color change took place when Solutions A and B were added, the only possibility left is that the bacteria reduced the nitrate to nitrogen gas.

4. *Urease production.* Pick up two plates of urea agar plates/slants/broths once all other inoculations are completed. They should be located in a covered container because the color indicator, phenol red, is light sensitive. Phenol red will appropriately be a yellow color in the presence of the somewhat acidic urea and fuschia pink in the presence of alkaline ammonia. Inoculate one plate/slant/broth with *E. coli* and the other with *Proteus mirabilis.* If placed on a plate, streak for isolated colonies. This test determines whether the bacterium inoculated on the plate produces an enzyme capable of catabolizing urea into ammonia and carbon dioxide, thus raising the pH. Once inoculated, wrap these plates in separate sheets of aluminum foil. (Why?)

Inventory

At the end of this exercise, each group will have inoculated:

Four tubes of Phenol Red Glucose broth containing Durham tubes
Four tubes of Phenol Red Lactose broth containing Durham tubes
Six tubes of O.F. Basal medium (remember to add the oil)
Four tubes of Simmons Citrate medium
Four tubes of Decarboxylase broth with lysine and/or
Four tubes of Decarboxylase broth with ornithine and/or
Four tubes of Decarboxylase broth with arginine (remember to add the oil)
Four tubes of T-Soy or Tryptone broth to test for indole production
Four tubes of Tryptic Nitrate medium to test for nitrate reduction
Two plates of urea Agar

Results

"Read" the reactions in the tubes and plates after incubation.

Decarboxylase Tubes

Reactions

Negative = yellow. The bacteria did not remove the carboxyl group from the amino acid tested. The tube is acidic.

Positive = purple. The enzyme decarboxylase was present for the amino acid in the tube. The pH of the tube is alkaline. Since the microbe may have partially utilized the pH indicator in the tube, any shade of purple is considered positive.

Indole Production. Add several drops of Kovac's reagent to each tube. *Do not mix.* Allow the reagent to stay concentrated on top of the broth.

Reminder: Be careful when using Kovac's reagent. It is toxic, irritating, and a suspected teratogen. Use the fume hood when using this reagent. If directed, use gloves also.

Reactions

Positive = red-colored ring. Indole was produced as a byproduct of tryptophan catabolism.

Negative = yellow. Indole was not produced by tryptophan catabolism.

Nitrate Reduction. Add several drops of Solutions A and B to each tube. *Do not shake.* Any tube showing color change indicates that *nitrite* is in that tube. Add *zinc* to any tube that shows no color change within five minutes.

Reactions

Positive for nitrite = red color with Solutions A and B

Positive for N_2 gas = no color change after zinc is added

Negative reaction = red color after zinc is added

Urease Production

Reactions

Positive = fuschia pink color indicates that urease catabolized urea to ammonia

Negative = no color change (plate remains salmon colored)

Note: There is usually no significant growth on these agar plates. The reaction is so sensitive that the bacteria placed on the plate contain enough enzymes to utilize the substrate and cause a positive reaction without growth.

After observing the decarboxylase, indole, nitrate reduction and urease reactions, fill out Table 11.2.

TABLE 11.2

NITROGEN METABOLISM REACTIONS

	E. coli	P. aerug.	P. mirab	B. subtilis
Lysine Decarb.				
Ornithine Decarb.				
Arginine Decarb.				
Indole				
NO$_3$ Reduction				
Urease		—		—

NAME _____ DATE _____ SECTION _____

QUESTIONS

Answer the following questions about the media inoculated in this exercise.

1. Phenol red tubes

   What color change do you see?

   Did any of the assigned microbes cause both tubes to change color?

   Was any microbe negative for the fermentation of both sugars?

   Did any of the Durham tubes with negative reactions show gas production?

2. O.F. basal medium

   Is the tube with the oil used to test for fermentation or for oxidation?

   What is the significance of having the tube without the oil change color and the tube with the oil remain green?

   What is the significance of having both tubes change color?

   What is the significance of having neither tube change color?

   Would it be possible for the tube with the oil to change color and the tube without the oil to remain green? Why or why not?

3. Simmons citrate agar

Exactly what does a positive reaction in a Simmons citrate tube indicate about the ability of the bacterium to utilize energy?

4. Decarboxylase broth

If more than one amino acid was used in this exercise, was any specific microbe positive for all reactions? Was it one enzyme or more than one that removed the carboxyl group from all amino acids?

Why was oil added to the tubes?

Did any tubes show up purple and clear (i.e., not cloudy)? Why would this affect your interpretation of the results?

5. Indole production

What precautions should be taken when performing this test?

What reagent is added?

6. Nitrate reduction

What reagents are added to test for this reaction?

What does zinc test for when added?

What does a red color indicate when zinc is added?

7. Urease test

Why is there still a reaction on the urease agar/broth, although there may be no significant growth?

MATCHING

a. turns blue when its substrate is utilized

b. used to detect the presence of gas formation

c. detects the presence of indole

d. red when alkali, yellow when acid

e. reduces nitrate to nitrite

f. enzyme that raises pH when its substrate (amino acid) is present

g. used to indicate whether nitrite is present

h. purple color indicates alkaline pH

i. enzyme that produces ammonia

j. turns yellow when glucose is oxidized

_____ O.F. basal medium

_____ Simmons citrate medium

_____ Kovac's reagent

_____ Zinc

_____ Solutions A and B

_____ Urease

_____ Durham tube

_____ Phenol red

_____ Decarboxylase

_____ Decarboxylase broth

MULTIPLE CHOICE

1. A phenol red tube was inoculated. Which substrate would be tested for?
   a. Kovac's reagent   b. nitrate   c. carbohydrate   d. amino acid

2. O.F. basal medium was inoculated. A layer of sterile mineral oil was then added. What reaction is tested for?
   a. tryptophan utilization   b. urea oxidation   c. nitrate reduction   d. carbohydrate fermentation

3. A green agar slant was inoculated. After 24 hours of incubation, much of the slant turned blue. Which is the true statement?
   a. It was used to test for citrate utilization.
   b. The amino acid in the tube lost its carboxyl group.
   c. Ammonia was a byproduct of the reaction.
   d. If no color change was seen, zinc would be added.

4. Solutions A and B turn red when added to a broth tube. The tube contains:
   a. urea   b. tryptophan   c. indole   d. nitrite

5. Urease breaks down urea to:
   a. nitrogen gas   b. indole   c. tryptophan   d. ammonia

6. A decarboxylase broth tube was inoculated with a microbe able to ferment glucose. The color of the tube after 12 hours of incubation would most likely be:
   a. yellow   b. red   c. green   d. blue

7. The chemical(s) used to detect the presence of indole from the breakdown of tryphophan is:
   a. Solutions A and B   b. Kovac's reagent   c. zinc   d. mineral oil

8. If a microbe produces gas when it utilizes a carbohydrate, the presence of this gas can be determined by:
   a. addition of zinc   b. odor   c. use of a Durham tube   d. addition of peroxide

WORKING DEFINITIONS AND TERMS

Aerobic Chemical reaction that requires the use of oxygen gas (oxidation).

Anaerobic Chemical reaction that does not utilize oxygen (fermentation).

Carboxylic acid Portion of an amino acid. When removed due to the action of the enzyme decarboxylase, the pH of the solution will rise. (Also called a carboxyl group.)

Catabolize Process of chemically breaking down larger molecules to smaller ones, usually for energy production.

Color indicator A substance that changes color at different pH levels or when a certain reagent is added.

Durham tube Small tube placed upside down in phenol red broth tubes. It is used to determine whether gas is produced as a byproduct of sugar fermentation.

Fermentation See anaerobic.

Read Term often used to describe the observing of a chemical reaction.

Substrate Any substance acted upon by an enzyme.

EXERCISE 12

Gas Requirements of Microorganisms

Objectives

After completing this lab, you should be able to:

1. Differentiate microbes according to their gas requirements.

2. Explain the various ways to grow anaerobic bacteria.

3. Describe the use of an anaerobic jar.

4. Explain why there is bubbling when hydrogen peroxide is placed on a wound.

Most of the bacteria used in basic microbiology laboratories grow well in, or at least tolerate, normal atmospheric concentrations of oxygen. The use of such microbes makes the preparation and maintenance of laboratory cultures easy and simplifies many of the procedures used in the lab itself. Certain bacteria have little or no tolerance for atmospheric oxygen. Growth of these bacteria requires modification of standard procedures already learned.

Most of the bacteria we use are *strict (obligate) aerobes* or *facultative anaerobes*. A strict aerobe requires close to a 21% oxygen atmosphere to survive. Place a typical strict aerobe in 0% oxygen and it will soon die. Facultative anaerobes are most often naturally found in or close to 0% oxygen atmosphere. Place them in normal atmospheric oxygen and they will grow just as well, if not better. This is because they contain enzymes that allow growth and survival in both high-oxygen (aerobic) and very low-oxygen conditions (anaerobic). When in a high-oxygen environment, they utilize it as part of their respiratory process, in a manner similar to human cells. When in an anaerobic environment, they utilize other chemicals for their metabolism by converting to a fermentative pathway. Thus, oxygen is not a limiting

factor for their growth. For example, *E. coli,* a common intestinal bacterium, is naturally found in an anaerobic environment but thrives on the surface of culture medium in a Petri dish.

The *microaerophiles* require between 5% and 15% atmospheric oxygen. Most tissues of our body have the equivalent of such a concentration. *Helicobacter pylori,* the causative agent of ulcers, and *Campylobacter jejuni,* which causes intestinal infections, are two such microaerophiles.

All of the microbes mentioned so far have one thing in common—enzymes that protect them from the poisonous aspects of oxygen. Oxygen usually acts as an electron acceptor in respiratory energy metabolism. It also gets involved in other cellular chemical reactions to produce *superoxides* and *peroxides*. These chemicals are highly reactive and can cause much damage if they are not quickly neutralized. Aerobic organisms, including humans, possess protective enzymes that successfully neutralize such compounds. *Strict anaerobes* have no such protective enzymes, and exposure to atmospheric oxygen soon kills most of them. *Treponema pallidum,* the causative agent of syphilis, dies within seconds when placed in an aerobic environment.

There is one group of bacteria that can exist in the presence of oxygen but do not use oxygen for metabolism. They are called *aerotolerant*. *Streptococcus pyogenes*, the causative agent of strep throat, is one such microbe.

Some bacteria also need somewhat higher levels of carbon dioxide than those found in the atmosphere. They require between 3% and 10% concentration, depending on the species. Once again, tissues of the human body provide such an environment. *Neisseria gonorrhea*, the causative agent of gonorrhea, is an example of a *capneic* microbe, or one that requires this higher level of carbon dioxide.

GROWTH OF ANAEROBES (may be done as a demonstration)

Materials

Waterbath

Boiling water

T-Soy broth

Sterile mineral oil

Melted agar deeps

Thioglycollate broth with color indicator

BBL GAS PAC™ System

Broth cultures of *E. coli*, *B. subtilis*, *Clostridium sporogenes*

PROCEDURE

E. coli is a well-known facultative anaerobe, *B. subtilis* is a strict aerobe, and *Clostridium sporogenes* is a spore-forming strict anaerobe. The presence of spores will allow *Clostridium sporogenes* to survive while being handled in a 21% oxygen atmosphere.

Boiled Broth. Boiling removes gases, such as oxygen, from liquid.

1. Take three tubes of freshly boiled and cooled T-soy broth.
2. Inoculate each with a different test bacterium. (Use a loop.)
3. Aseptically add 2 to 3 mls of sterile mineral oil to each tube so that anaerobic conditions are maintained.

Melted Agar. A temperature of 100° C is required to melt agar. Melted agar therefore is free of atmospheric oxygen.

1. Allow 4 melted T-soy agar deeps to cool in a water-bath maintained at 45° C.
2. Use a loop to inoculate the three tubes with the test bacteria.
3. Cool rapidly by placing the melted agar deeps in a beaker of cold water. Once the agar has solidified, sterile mineral oil does not have to be added because solidified agar prevents the penetration of oxygen.
4. Place an uninoculated tube with those inoculated. This control tube will later be used to compare evidence of growth by measuring cloudiness within the agar.

Note: Make sure all the tubes are the same color when they are inoculated. Different shades of the same color will make it difficult to determine whether growth did occur.

Thioglycollate Broth. Thioglycollate acts as a reducing agent for oxygen. As oxygen penetrates the broth, it reacts with the thioglycollate, preventing it from interfering with anaerobic metabolism. Many preparations of thioglycollate contain a color indicator for oxygen saturation. The presence of a light green or red color in the upper part of the tube indicates that oxygenation has taken place.

1. Inoculate three tubes of thioglycollate broth with the assigned bacteria.
2. Use a loop and insert it throughout the entire tube so that the inoculum is inserted into anaerobic as well as aerobic regions of the broth.

Anaerobic Jar Method. All the methods used so far have been simple, effective ways to grow anaerobes, but these methods lack the ability to produce isolated colonies. Anaerobic incubators and inoculating boxes, sometimes costing thousands of dollars, can accomplish this task of isolating even the most oxygen-sensitive anaerobe. A less expensive alternative method is to use a container that has the oxygen, or at least most of the oxygen, removed. Originally, the process involved placing a lit candle within a wide-mouthed jar with freshly inoculated agar plates. The lid was then tightened. The apparatus was appropriately called a *candle jar*. This procedure not only reduced the amount of oxygen to very low levels, but also raised the carbon dioxide levels for capneic microbes. A more modern method involves using a commercially available anaerobic jar such as that utilized by the Baltimore Biological Laboratory (BBL)™. (See Fig. 12.1.) It uses a wide-mouthed container into which inoculated plates are placed. A disposable hydrogen generator packet is placed in the con-

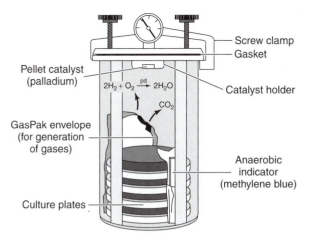

FIG. 12.1. Diagram of BBL Gas Pak™.

tainer along with a catalyst, which allows the hydrogen to react with any oxygen, resulting in the production of water. (Carbon dioxide is also a byproduct of the reaction packet used in this procedure.) Streak out 2 plates each of the 3 test bacteria. Place one set in the anaerobic jar and allow the other set to grow under aerobic conditions.

Results

Broth Tubes. Observe the T-soy broth tubes with oil. Shake gently. Which tube or tubes show growth? Record your results.

	Growth or No Growth
Bacillus subtilis	
E. coli	
Cl. sporogenes	

Agar Deeps. Look for evidence of growth in the agar deeps. Compare the tubes with the uninoculated control tube if necessary.

	Growth or No Growth
Bacillus subtilis	
E. coli	
Cl. sporogenes	

Thioglycollate Broth. Do not shake. Observe these tubes. Where did growth occur within each tube?

	Top of Tube	Throughout the Tube	Bottom of Tube
Bacillus subtilis			
E. coli			
Cl. sporogenes			

Gas Pac. Compare the growth of the bacteria from agar plates incubated in the BBL Gas Pac™ system with those grown in atmospheric oxygen. Which microbes grew in both conditions?

	Growth in Gas Pac System	Growth in Atmospheric Oxygen
Bacillus subtilis		
E. coli		
Cl. sporogenes		

DEMONSTRATION OF CATALASE

Catalase is an enzyme capable of breaking down hydrogen peroxide, a common byproduct of oxygen metabolism, into water and oxygen. Aerobic and facultative bacterial cells, as well as many human cells, contain this enzyme. Red blood cells contain especially high levels of catalase since their mission is to transport oxygen to body tissues.

Materials

3% hydrogen peroxide solution

Dropper

Agar plate cultures of *Enterococcus faecalis, E. coli, Clostridium sporogenes, B. subtilis* (If the procedure is done after incubation of the anaerobic jar method, only a plate of *E. faecalis* has to be provided.)

PROCEDURE

Open the agar plates and add 1–2 drops of hydrogen peroxide to each of the bacteria growing there. Record which one(s) showed bubbling indicative of oxygen production and which one(s) did not. The bubbling is indicative of the presence of catalase.

NAME _____ DATE _____ SECTION _____

QUESTIONS

1. Where would a microaerophile grow in thioglycollate broth?

2. The BBL Gas Pac™ showed that an anaerobe such as the spore-forming *Clostridium sporogenes* could be culti-vated in a microbiology laboratory without specialized transferring equipment. Why couldn't the same procedure be used to prepare obligate anaerobic vegetative cells?

3. *Enterococcus faecalis* showed no bubbling in the catalase test, yet it grew on the surface of an agar plate outside of the BBL Gas Pac™. What kind of oxygen requirement would this microbe have?

MATCHING

a. aerotolerant

b. facultative anaerobe

c. capneic

d. catalase

e. strict (obligate) anaerobe

f. thioglycollate

g. microaerophile

_____ requires high concentrations (3–10%) carbon dioxide for optimum growth

_____ reacts with free oxygen, thus removing it from a solution

_____ requires an oxygen-free environment to grow

_____ has enzymes to allow growth in both high-oxygen and low-oxygen envi-ronments

_____ converts hydrogen peroxide into oxygen and water

_____ does not use oxygen for its metabolism and is not harmed by its presence

_____ requires 5–15% oxygen environment for optimum growth

MULTIPLE CHOICE

1. A microbe that requires close to atmospheric oxygen for adequate growth would be considered:

    a. anaerobic   b. aerobic   c. a facultative anaerobe   d. capneic

2. A microbe that requires higher than atmospheric carbon dioxide for growth is:

    a. anaerobic   b. aerobic   c. facultative anaerobe   d. capneic

3. Hydrogen peroxide is placed on a wound, and blood is present. The bubbling seen is due to:

    a. catalase   b. thioglycollate   c. a facultative anaerobe   d. clotting factors

4. A facultative anaerobe is inoculated in a tube of thioglycollate broth. Where would you expect to see growth?

    a. top   b. middle   c. bottom   d. all of these

5. An aerotolerant anaerobe is inoculated in a tube of thioglycollate broth. Where would you expect to see growth?

    a. top   b. middle   c. bottom   d. all of these

6. A microaerophile is inoculated in a tube of thioglycollate broth. Where would you expect to see growth?

    a. top   b. middle   c. bottom   d. all of these

7. A microbe that has two sets of enzymes to allow it to grow in both oxygen-rich and oxygen-free environments is:

    a. anaerobic   b. aerobic   c. a facultative anaerobe   d. capneic

8. A microbe that can grow in the presence of oxygen but does not use it in its metabolism is:

    a. capneic   b. aerotolerant   c. a facultative anaerobe   d. microaerophilic

WORKING DEFINITIONS AND TERMS

Aerotolerant   A microbe that can grow in the presence of atmospheric oxygen but does not use it in their metabolism.

Candle jar   A device that uses a lit candle to reduce the concentration of atmospheric oxygen as well as increase the amount of carbon dioxide in the container.

Capneic   An organism that requires a higher than atmospheric concentration of carbon dioxide (usually 3–10%).

Catalase   A protective enzyme used by aerobic organisms, which breaks down hydrogen peroxide, a toxic byproduct of oxygen metabolism.

Facultative anaerobe   A microbe that can exist in both the presence and absence of oxygen. If oxygen is present, it has enzymes to utilize it as part of its respiratory metabolism. If oxygen is not present, it utilizes other enzymes for its energy metabolism.

Microaerophilic   A microbe that requires lower than atmospheric levels of oxygen to grow.

Strict (obligate) aerobe   A microbe that requires atmospheric levels of oxygen to survive.

Strict (obligate) anaerobe   A microbe that requires an oxygen-free environment to survive.

PROCEDURE: AGAR SLANTS

Note that many of the bacteria provided for you and the ones you inoculated in Exercise 2 can be differentiated from each other on the basis of the type of growth on the agar slant. Observe Fig. 3.2, which shows some of the different forms of growth seen on agar slants, and compare this with the types of growth seen in the tubes at your table or workstation.

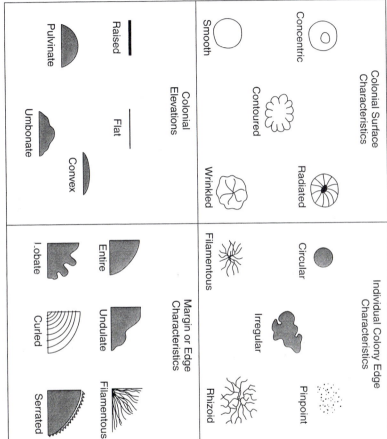

FIG. 3.1. Colonial characteristics.

Colonial Surface Characteristics				Individual Colony Edge Characteristics		
Concentric	Contoured	Radiated	Wrinkled	Circular	Irregular	Pinpoint
Smooth				Filamentous	Rhizoid	
Colonial Elevations				Margin or Edge Characteristics		
Raised	Flat	Convex		Entire	Undulate	Filamentous
Pulvinate	Umbonate			Lobate	Curled	Serrated

PROCEDURE: BROTH

Before observing the growth characteristics in the broth tubes at your table, review the types of growth seen in Fig. 3.3. Unless you have a fresh, 24-hour culture right from the incubator, you will not see many tubes with turbidity. This is because the bacteria tend to settle out or to precipitate when they are refrigerated for several days. After you note that either a *pellicle* or *ring for-*

Specialized Media

Objectives

After completing this lab, you should be able to:

1. Explain the function of enriched, selective, highly selective, differential, and multitest media.

2. Differentiate between alpha, beta, and gamma hemolysis reactions on blood agar.

3. Describe the function of MacConkey (MAC) and Eosin-Methlyene Blue (EMB) agar.

4. Explain the function of Mannitol Salt Agar (MSA).

5. Explain the function of Phenyl Ethyl Alcohol (PEA) Agar.

6. Describe the possible reactions of Triple Sugar Iron Agar (TSIA) and Sulfide-Indole-Motility (SIM) medium.

Medical microbiology requires speed and accuracy in the identification of bacteria. Often a delay of even one day in the laboratory identification of a microbe may result in serious consequences for a patient. For this reason, various shortcuts have been developed that lead to rapid identification of a suspected pathogen without the loss of accuracy. Specialized media have aided in this quest.

Until now, you have been using general-purpose media for bacterial growth. Though adequate for ordinary laboratory procedures, modifications must be made for labs specializing in areas such as medical microbiology, food and water investigations, and environmental studies.

Enriched media contain extra or higher concentrations of nutrients, vitamins, trace elements, and other growth factors that allow *fastidious* microbes to grow under laboratory conditions. Such microbes may grow quite easily in a human host where all of these ingredients are readily available but will not grow when placed on ordinary or standard growth media. Blood contains many of these needed growth factors and, when added to growth media, allows many of these fastidious organisms to flourish in the laboratory. Such media are

needed in the microbiology laboratory because if you can't grow it, you can't identify it. (This last statement is not entirely true as advances in serology continue.)

Differential media is one such category of media meant to speed up the identification of a microbe Any medium that allows *all* microbes to grow in such a way as to allow them to be distinguished or categorized by their growth would be considered differential. Based on this statement, all the media used in Exercise 11 are differential as the bacteria can now be classified as citrate $(+)$ or $(-)$, indole $(+)$ or $(-)$ and so on.

Selective media have chemicals added that will inhibit the growth of some microbes but not others. Thus, certain microbes will grow on this type of medium but not others. For example, the addition of penicillin to a growth medium will *select* for the growth of penicillin-resistant bacteria and inhibit the growth of bacteria sensitive to penicillin.

Highly selective media are a type of selective media that contain chemical formulas which inhibit almost every microbe except one genus and/or a few species. For example, the *Staphylococci* are the only human flora that will grow well on a salt concentration greater than 7%. If such a high concentration of salt is placed on an agar plate (actually 7.5%), the only bacteria from humans that would be isolated on this medium would be this genus.

Multitest media or *combination differential media,* as the name implies, perform several different biochemical tests with only a single inoculation. You may recall that in Exercise 11, two phenol red broth tubes were needed to determine whether a microbe had the enzymes to ferment glucose and lactose. A single inoculation into *Triple Sugar Iron Agar* can make the same determination as well as test for two other types of reactions.

Selective and *differential media* contain chemicals that will allow certain microbes to grow while preventing others from growing (selective), as well as causing the ones that grow to appear "different" so they can be distinguished from each other.

EXAMPLES

Blood Agar Plates (BAP) are considered a type of *differential media.* Virtually all types of bacteria associated with humans will grow on it, but they will grow in three distinct ways. If they release enzymes, called *hemolysins,* which are **partially** able to destroy red blood cells, the area *adjacent* to the bacterial growth will appear green when the plate is held up to the light. This indicates that the microbe is alpha (α)-hemolytic or shows α-hemolysis. If their enzymes **completely** destroy red blood cells, the area adjacent to the bacterial growth will appear clear. This type of microbe is termed beta (β)-hemolytic (β-hemolysis). If **no enzymes** are present for red blood cell destruction, the adjacent area remains red, and this phenomenon is termed gamma(γ)-hemolysis, or the microbe is considered γ-hemolytic or nonhemolytic.

Phenyl-Ethyl Alcohol (PEA) agar is able to dissolve the lipopolysaccharides of Gram negative bacteria. Gram negatives are thus inhibited while Gram positives are allowed to grow. PEA agar is therefore an example of a *selective medium.*

Mannitol Salt Agar (MSA) contains 7.5% salt. The only type of microbes normally found associated with the human body, capable of growing well in this environment, are the staphylococci. *Staphylococcus aureus* also has the ability to ferment the sugar mannitol, which is present in this medium. Fermentation of the mannitol triggers a pH change, which then causes a color change. Since the *Staphylococci* are the only microbes that will grow well on this medium, Mannitol Salt Agar is considered *highly selective* for this genus. In addition, the mannitol enables one to differentiate between the highly pathogenic *S. aureus* and other, less dangerous members of this genus, such as nonmannitol-fermenting *S. epidermidis.*

Selective and *differential* media are exemplified by MacConkey (MAC) and *Eosin-Methylene Blue* (EMB) agars. Both contain chemicals that inhibit the growth of Gram positive bacteria and allow for the growth of the Gram negatives. Both also contain the carbohydrate lactose, along with a pH color indicator. You can, therefore, "weed out" any Gram positives mixed with Gram negatives. Growth on either of these two media will indicate that the test microbe is Gram negative and whether or not it ferments lactose (color change along with the growth). *MacConkey agar* has the advantage of completely inhibiting the growth of Gram positives (no growth at all) but can sometimes inhibit some Gram negatives. *Eosin-Methylene Blue* does allow some Gram positives to grow in a limited way but has the advantage of often showing the lactose-fermenting ability of *E. coli* and *Klebsiella pneumoniae* by the presence of metallic green colonies.

Triple Sugar Iron Agar (TSIA) and *Sulfide-Indole-Motility* (SIM) are examples of *multitest media.* One inoculation in each of these will test for several different reactions. *Triple Sugar Iron Agar* is a slant that contains glucose, lactose, and sucrose. For the purpose of simplification, we will ignore sucrose in the following discussion. The tube will be inoculated with a needle so that bacteria are growing within the agar (anaerobic fermentation), as well as on the surface of the slant (aerobic oxidation). The fermentation of the glucose will cause enough of a pH change to trigger a color change in the butt, or bottom, of the tube from red to yellow. Fermentation of glucose *and* lactose will result in the butt *and* the slant turning yellow. The *iron* listed in the title of the medium is an iron salt with sulfide as a component. If iron is utilized by the bacteria, hydrogen sulfide (H_2S) will be left as a byproduct, turning the medium black. Finally, if CO_2 gas is produced as a result of fermentation, cracks or bubbles will be observed in the agar. In other words, one inoculation will determine glucose fermentation, glucose plus lactose fermentation, H_2S production, and whether gas was produced from glucose fermentation. A deeper red color to the agar slant indicates an alkaline reaction.

Sulfide-Indole-Motility (SIM) is also a *multitest medium* as a single inoculation will test for three phenomena:

S = sulfide = H_2S production

I = indole production

M = motility

Sulfide production will turn the tube black. Indole will show up as a red color once Kovacs reagent is added. Motility will show up as a "cloudy" region in the semi-solid growth medium because motile bacteria are able to move away from the inoculation site. Nonmotile bacteria will show up as growth only along the line of inoculation.

INOCULATION OF BLOOD, PHENYL-ETHYL ALCOHOL, MANNITOL SALT, MACCONKEY, AND EOSIN-METHYLENE BLUE AGAR PLATES

Materials List Per Table/Workstation

Broth cultures of: *Bacillus subtilis or cereus, Escherichia coli, Pseudomonas aeruginosa, Serratia marcescens, Staphylococcus aureus, Staphylococcus epidermitis*

Agar plate culture of: *Proteus mirabilis*

Four Blood Agar Plates

Three each of Phenyl-Ethyl Alcohol, Mannitol Salt MacConkey and Eosin-Methylene Blue Plates

PROCEDURE

1. Pick up and label the plates listed above. Make sure they are properly labeled. Do not rely on remembering their colors for these colors will change once growth occurs.

2. Take one Blood Agar Plate and inoculate it as follows:

   a. Take a sterile cotton swab and touch the back of your throat (see Fig. 13.1).

   b. Streak the swab over the surface of the plate, leaving much space between the streak marks. This procedure is necessary as pharyngeal bacteria will

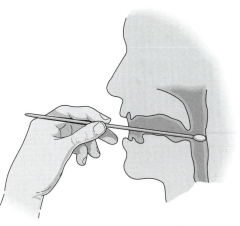

FIG. 13.1. Throat culture procedure.

FIG. 13.2. Make sure *Proteus mirabilis* is always placed on a separate plate.

show excellent alpha (α) hemolysis, whereas the stock bacteria provided each week tend to attenuate and lose their ability to demonstrate this form of hemolysis.

3. Inoculate one each of the Blood Agar, Phenyl-Ethyl-Alcohol, Mannitol Salt, MacConkey, and Eosin-Methylene Blue plates with *Proteus mirabilis,* with the streak technique for achieving isolated colonies. This particular microbe produces numerous flagella, which often enables it to *swarm* or spread over certain types of media. (See Fig. 13.2).

4. Divide the other two plates of each medium into three sections and pre-label with the name of each of the assigned bacteria. You now have a total of six sections available for bacterial growth. Place a sample of each of the other six bacteria in each section, leaving space between each inoculation. (See Fig. 13.2). You may be directed to use a cotton swab in the same manner as was done in the inoculation of the ultra-violet light plates. (See Exercise 10.)*

Results

Blood Agar Plate (BAP). Note that all the bacteria grew. Determine whether swarming occurred on any of the plates. Hold up the plates to the light and determine the type of hemolysis *adjacent* to the growth.

> α = green zone
>
> β = clear zone
>
> γ = no color change (nonhemolytic)

Phenyl Ethyl Alcohol (PEA) Agar. Determine which cultures grew well and which ones were completely inhibited or displayed impaired growth. (Compare the growth on this plate with that on the Blood Agar Plate to determine whether the growth is indeed impaired.)

---

*When you inoculate the broth cultures onto these plates, you will achieve more accurate results if you dilute the inoculum first. Otherwise, the higher concentration of bacteria placed on these plates may overcome any inhibitory chemicals used to prevent their growth. You can easily dilute the inoculum by placing a sterile swab into the broth tube of bacteria and then transferring that swab into a tube of sterile saline solution. Once done, you can use the same swab or a loop to place the diluted sample of bacteria onto each plate.

TABLE 13.1 — GROWTH CHARACTERISTICS ON INOCULATED AGAR PLATES

Medium	B. sub or cereus	E. coli	Ps. aerug.	Serr. marc.	Staph. aureus.	Staph. epid.	Prot. mir.
BAP hemolysis							
PEA							
MSA							
MAC							
EMB							

Growth = Gram positive organism

No growth or impaired growth = Gram negative organism

Mannitol Salt Agar (MSA). Note that only the *Staphylococcus aureus* grew well and fermented the mannitol.

Growth = *Staphylococcus spp.*

Growth + color change = *Staphylococcus aureus*

MacConkey (MAC) Agar. Determine which cultures grew well and which ones did not grow at all. Note the pink color which indicates lactose fermentation.

Growth = Gram negative organism

Growth + color change of colony to pink = Gram negative and lactose positive

Eosin-Methylene Blue (EMB) Agar. Determine which cultures grew well and which ones did not grow at all or displayed impaired growth. (Compare the growth on this plate with that on the Blood Agar Plate to determine whether the growth is indeed impaired.) Note that the purple color of the growth indicates lactose fermentation. Ascertain if *E. coli* grew in the characteristic green sheen mentioned previously.

Fill in Table 13.1 after observing the growth characteristics on the inoculated agar plates.

INOCULATION OF TRIPLE SUGAR IRON AGAR AND SULFIDE-INDOLE-MOTILITY MEDIUM

Materials List Per Table/Workstation

Divided agar plate of *Escherichia coli, Pseudomonas aeruginosa, Serratia marcescens*

Agar plate of *Proteus mirabilis*

Four tubes of Triple Sugar Iron Agar

Four tubes of Sulfide-Indole-Motility Medium

PROCEDURES

1. Inoculate the four tubes of Triple Sugar Iron Agar (TSIA) with the assigned bacteria using an inoculating needle. When inoculating each tube, push the needle *into* the bottom of the agar slant (butt), withdraw it, and streak the slant. When completed, each tube will have bacteria growing within the agar (anaerobic conditions) and on the surface (aerobic conditions). (See Fig. 13.3.)

2. Inoculate the four tubes of Sulfide-Indole-Motility (SIM) medium with the assigned bacteria using a needle or stab. Penetrate each tube approximately halfway down into the medium. Try to avoid moving the needle too much to either side. (See Fig. 13.4.)

FIG. 13.3. Inoculation of TSIA.

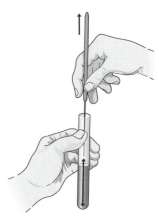

FIG. 13.4. Inoculation of SIM.

TABLE 13.2

REACTIONS OBSERVED ON TSIA AND SIM MEDIA

Microbe	Triple Sugar Iron Agar				Sulfide Indole Motility Medium		
	Slant Reaction	*Butt Reaction*	*CO_2 Gas Production*	*H_2S Production*	*H_2S Production*	*Indole*	*Motility*
E. coli							
Pseudomonas aeruginosa							
Serratia marcescens							
Proteus mirabilis							

(Use the same technique as for the O.F. basal medium inoculation in Exercise 11.)

Results

Triple Sugar Iron Agar. Examine the TSIA tubes and determine whether glucose and/or lactose were fermented, whether hydrogen sulfide (H_2S), was produced, and whether gas was produced from the fermentation of glucose.

Glucose fermentation = yellow butt and red slant

Glucose and lactose fermentation
= yellow butt and yellow slant

Hydrogen sulfide production = blackening of the agar

Gas production = cracks/bubbles in the agar

Sulfide-Indole-Motility Medium. Examine the SIM medium tubes and determine whether hydrogen sulfide was produced. Hold up the tubes to the light and examine them for evidence of motility. Nonmotile bacteria will appear as sharp growth only along the line of inoculation. Motility will appear "cloudy" or "fuzzy," for the microbes were able to move away from the stab line of inoculation. Add approximately 10 drops of Kovacs reagent (**Caution: use hood**) to read the indole reaction. (If *S. marcescens* grew with a red color, this reaction might be difficult to read in this tube with this microbe.)

Sulfide = blackening of the tube

Indole = red color after the addition of Kovacs reagent

Motility = nondistinct line of growth in the medium or cloudiness in the medium

Inventory

At the end of this exercise, each group will have inoculated:

Four Blood Agar plates
Three Phenyl-Ethyl Alcohol Agar plates
Three Mannitol Salt Agar plates
Three MacConkey agar plates
Three Eosin-Methylene Blue agar plates
Four Triple Sugar Iron Agar tubes
Four Sulfide-Indole-Motility medium tubes

MEDIA SUMMARY—EXERCISES 11 AND 13

Medium	Type[1]	Reaction Observed[2]	Positive/Negative[3]	Reagent(s) Used or Present[4]	Specific Microbe Identified[5]
Phenol red broth					
O. F. basal					
Simmons citrate					
Decarb. medium					
Indole					
Nitrate reduct.					
Urease					
BAP					
PEA					
MSA					
MAC					
EMB					
TSIA					
SIM					

[1]**TYPE:** selective, differential, highly selective, both selective and differential.
[2]**REACTION OBSERVED:** what reaction took place, what chemical was utilized or produced, etc?
[3]**+/-:** what color, growth pattern, etc. determines a positive or a negative reaction.
[4]**REAGENT(S):** what chemicals have to be added after growth?
[5]**SPECIFIC MICROBE:** for media that is highly selective. Which microbe is the medium used to isolate or identify?

NAME _____ DATE _____ SECTION _____

QUESTIONS

1. What is the difference between differential and selective media?

2. How is alpha and beta hemolysis recognized on blood agar? Why is the term *gamma hemolysis* actually a misnomer?

3. How many different reactions can be seen in a TSIA tube?

4. How many different reactions can be seen in a SIM tube?

5. Fill in the chart on the previous page: MEDIA SUMMARY.

MATCHING

a. *Proteus mirabilis*

b. *E. coli*

c. Mannitol Salt Agar

d. Phenyl Ethyl Alcohol agar

e. differential medium

f. multitest medium

g. phenol red broth, O.F. basal medium, decarboxylase broth, urease test agar

h. MacConkey agar

i. Eosin-Methylene Blue agar, MacConkey agar, Phenyl Ethyl Alcohol agar

_____ highly selective for staphylococci

_____ most likely to grow with a green metallic sheen on EMB

_____ all microbes grow on this medium, and the type of growth allows us to classify them

_____ one inoculation will test for several biochemical reactions

_____ examples of differential media

_____ most likely to swarm

_____ inhibits Gram (+) and differentiates Gram (−) into lactose positive and negative

_____ contains chemicals that dissolve lipopolysaccharides

MULTIPLE CHOICE

1. An example of a differential medium is:

   a. blood agar plate   b. decarboxylase broth   c. tryptic nitrate broth   d. all of these

2. A microbe grows with a green halo on a blood agar plate. This is an indication:

   a. that indole was not produced   b. that citrate was not utilized

   c. of lactose fermentation   d. of alpha hemolysis

3. An agar growth plate was prepared so the final pH was 5.0 rather than the usual 7.0. This medium would most likely be considered:

   a. multiple test   b. selective

   c. differential   d. selective and differential

4. A microbe grows with a green metallic sheen on eosin methylene blue agar. Which statement is true?

   a. it is gram negative

   b. it rapidly ferments lactose

   c. it has a good likelihood of being *E. coli* or *Klebsiella pneumoniae*

   d. all of these

5. A bacterium is inoculated into a triple sugar iron agar slant. A blackening occurs in the tube. This indicates:

   a. lactose was fermented   b. the microbe is motile

   c. an iron salt was metabilized   d. indole was produced

6. *Staphylococcus aureus* is a Gram positive coccus. In which of the following would it not be expected to grow?

   a. mannitol salt agar   b. phenyl ethyl alcohol agar   c. MacConkey agar   d. blood agar plate

7. A bacterium was inoculated into triple sugar iron agar. After 24 hours, the butt and the slant were yellow, and there were cracks in the agar. What are the reactions?

   a. glucose (+), lactose (+), gas (+)   b. glucose (+), gas (+)

   c. nitrate reduction (+), motility (+)   d. glucose (−), lactose (+), gas (+)

8. *Salmonella typhi* is a gram negative, lactose negative bacillus that causes severe intestinal disease. It is often found in fecal material along with *E. coli* and numerous other types of bacteria. Which of the following would effectively isolate and help identify this pathogen from within a fecal sample?

   a. MacConkey agar   b. blood agar plate

   c. phenyl ethyl alcohol agar   d. mannitol salt agar

WORKING DEFINITIONS AND TERMS

Differential medium Medium that allows all microbes to grow on it, but the type of growth allows the observer to distinguish between various bacterial types.

Hemolysin Enzyme capable of breaking down red blood cells.

Hemolysis The lysing of red blood cells. Bacteria can be differentiated by how well they are able to lyse red blood cells (α,β,γ).

Highly selective medium Growth medium composed of chemicals that inhibit all but a few groups of mi-

crobes. Growth on such a medium virtually identifies the microbe that the medium is designed to test for.

Motility Ability of a microbe to move (proof of presence of flagella) from the initial point of inoculaton.

Multitest media Growth medium formulated to demonstrate several different aspects of bacterial metabolism within the same tube or plate.

Selective medium Growth medium that allows the growth of certain microbes and prevents or inhibits the growth of others.

IV MEDICAL MICROBIOLOGY

A priority of the clinical microbiologist is to identify suspected infectious material from patients both quickly and accurately. Delay in such identification may be detrimental and even fatal to patients who need definitive treatment. Over the last century, numerous techniques were developed to rapidly collect, process, identify, and determine the drug sensitivities of microbes, primarily bacteria. The overall process is often called performing a C & S, or **Culture and Sensitivity.** One such method of determining microbial sensitivity is the Kirby-Bauer procedure, which was addressed in the exercise on microbial control (Exercise 10).

The most modern methods include combining procedures of identification and determining drug sensitivities using automated, computerized equipment costing many thousands of dollars. Since it is highly unlikely that a college microbiology laboratory has such exotic equipment available, you will be introduced to the more traditional identification methods as well as somewhat more modern but less automated (and expensive) processes.

EXERCISE 14 Genetics

Objectives

After completing this lab, you should be able to:

1. Explain how an autotrophic bacterium can be utilized to demonstrate the presence of mutagenic chemicals.

2. Determine how to recognize mutations in ultraviolet irradiated *Serratia marcescens.*

3. Recognize the presence of antibiotic-resistant spontaneous mutations.

Our environment is filled with chemicals that are capable of causing mutations. In addition to these chemicals, physical aspects of our environment, such as ultraviolet (UV) light, also damage DNA, resulting in mutations. (Proteins are also severely damaged by this UV light—consider the effects overexposure to sunlight has on light skin.)

With literally thousands of chemicals finding their way into our food and water, spewing throughout our atmosphere, and winding up on environmental surfaces, it is important to have a low-cost screening procedure to determine whether any of these chemicals are capable of damaging DNA. Such chemicals may be potential *mutagenic* agents. The *Ames test,* named after Dr. Bruce Ames, is one such screening test. A special strain of *Salmonella typhimurium* is used to help determine whether such chemicals are part of our environment. Although all *Salmonella* species are considered pathogenic and grow readily on most standard microbiological growth media, this strain cannot grow on such media because it is missing the gene needed to synthesize the amino acid, *histidine.* If placed in a histidine-rich growth medium, it will grow as well as all other *Salmonella* strains. This deficient strain of *Salmonella* is known as an *auxotroph,* that is, an organism that has lost the ability to synthesize a needed substance and thus requires this substance in its environment. Remove the sub-

stance, and you remove the ability of the organism to grow.*

The strain of *S. typhimurium* used in the Ames test is known as "his⁻". If inoculated onto a medium deficient in histidine, any colonial growth seen would represent a *spontaneous mutation* back to a "his⁺" strain. That is, a *backmutation* occurred, and the microbe now has a working gene that allows it to produce its own histidine. If this "his⁻" strain is placed on the same type of histidine-deficient environment along with another chemical, and a large level of colonial growth is observed (above the rate of spontaneous mutation), this indicates that the other chemical is a mutagen. This type of mutation would be considered an *induced mutation.* In general, the stronger the mutagenic agent, the greater the mutation rate, and the greater the amount of colonial growth seen.

UV radiation also damages DNA. The genes that control pigment in *Serratia marcescens* are particularly sensitive to certain wavelengths of UV light. When exposed to UV light for appropriate amounts of time, these mutations become readily apparent by the observation of different colored colonies (e.g., white, pink, or orange).

---

*Based on this definition, humans may also be considered auxotrophs. We do not have the ability to produce many of our amino acids and vitamins required for survival.

Antibiotic resistance in bacteria is a constant concern in the clinical area. It is imperative that laboratory technicians be vigilant in the detection of these mutants. The Kirby-Bauer plate technique measures the effects of various antimicrobial drugs on specific bacteria. By covering the surface of the test plate with the test bacterium, placing a paper disk saturated with a standardized concentration of drug on the plate, and observing a "zone of inhibition" that surrounds each disk, sensitivity to specific drugs can be determined. If a colony does show up with one of these zones, and there was no contamination, the colony represents a mutation that renders that particular microbe resistant to that particular drug.

Materials List per Table/Work Station

Broth culture of *Salmonella typhimurium* (strain TA1538) ATCCe 29631: *Serratia marcescens, Staphylococcus aureus*

Four plates of glucose (-) minimal agar

Four melted tubes of 4 ml soft agar (glucose (-) minimal salts with 0.5 ml mM histidine and 0.5 ml mM of biotin kept in a water bath at 45°C

Test materials for Ames test: disks impregnated with the following materials—2-nitro fluorene/alcohol mixture; #2 Red dye from Marachino cherries; 1% phenol solution; hair dye—any brand; ½–1% sodium benzoate or benzoic acid; any disinfectants available (see Exercise 10); any liquid cosmetics. Also available for testing—thin slices of hot dogs; thin slices of used cigarette filters

Three Mueller-Hinton agar plates

Seven Nutrient Agar plates

UV light

Antibiotic disk dispenser

Sterile pipettes

AMES TEST

In this procedure, you will determine whether certain chemicals are capable of causing mutations. Such chemicals are called *mutagens*. When the auxotrophic strain of his⁻ *S. typhmurium* is exposed to a test chemical (placed on a paper disk), and a large amount of colonial growth is observed surrounding this disk, it becomes evident that this chemical is indeed a mutagen. The mutagenic chemical converted the his⁻ strain of the bacterium to a his⁺ strain (backmutation). The growth medium utilized in this procedure contains a very small amount of histidine plus another growth factor that will allow for some growth of the test organism, but not enough to completely cloud the plate with a typical "lawn" of bacteria. The medium used is thus termed *minimal growth medium* (See Figs. 14.1 and 14.2).

There are several variations of the Ames test. The following is one of the more basic procedures utilized. (More sophisticated versions include the use of liver extract to enhance the activity of the mutagens.)

1. Aseptically add 0.1 ml of the *S. typhimurium* broth to one of four melted soft agar deeps with the histidine/biotin mixture. *Work quickly before the mixture solidifies*. Mix well.

⚠ **CAUTION:** DO NOT PIPETTE BY MOUTH.

2. Pour the melted agar mixture over the minimal growth medium agar plate. Allow it to solidify.

3. Repeat steps 1 and 2 with the other three melted agar deeps and plates.

4. Aseptically place four sterile filter disks in the four divided sections of one of the plates. This will act as a control and determine how many spontaneous mutations are present. (See Fig. 14.1.)

5. Place various chemical-saturated disks in the remaining three plates (up to four per plate). Label each

FIG. 14.1. Minimal growth medium with a sterile disks showing random spontaneous mutations (control plate)

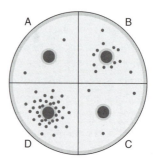

FIG. 14.2. Minimal growth medium with disks saturated with various chemicals. Note that disks A and C are non-mutagenic, disk B is slightly mutagenic, and disk D is highly mutagenic.

section so you know which disk contains which chemical.

6. Invert the plates and place in the incubation tray.

UV LIGHT PROCEDURE

Refer to Exercise 10 and follow the procedure covered there. If directed, completely cover the surface of the Nutrient Agar plate with a lawn of *S. marcescens* before exposure to the UV light.

KIRBY-BAUER PROCEDURE

Refer to Exercise 10 and follow the procedure covered there.

Results: Ames Test

Compare the control plate with the sterile disk showing spontaneous mutations (if any) with the other plates containing disks saturated with test chemicals. Compare the number and concentration of colonies surrounding the disks. Determine which chemicals are strongly mutagenic, moderately mutagenic, and nonmutagenic. Subtract the average number of colonies found in one-fourth of the control plate with that in each quarter of the test plates. For example, if the average number of spontaneous mutations in one-fourth of the control plate is 15, and the number of colonies in one of the test plate quarters is 40, the number of mutagenic induced mutations is $40 - 15 = 25$. More than 100 colonies in one-fourth of the test plates

(almost solid growth) indicates that the chemical is strongly mutagenic; if the number of colonies is between 10 and 100, it is moderately mutagenic; and if it has less than 10 colonies, it is slightly mutagenic. No difference in colonies indicates that the chemical is nonmutagenic. After observing your results, fill in the following chart.

Results: UV Light Procedure

Observe the growth patterns of *S. marcescens*. Determine which plate has the highest number of mutations and the amount of exposure that achieves this mutation rate (time versus distance versus intensity of the light source).

Results: Kirby-Bauer Test

Observe the zones of inhibition of the antimicrobial drugs on the *S. aureus* test plate (or any other test plate done). Look for colonies within the zone of inhibition. If instructed, perform additional tests to determine whether the colonies are the same genus and species of the test bacteria or whether they are the result of accidental contamination.

Inventory

At the end of this exercise, each table or work group will have completed:

Four Ames test agar plates
Seven Nutrient Agar plates covered with *S. marcescens* and exposed to UV light
One (or more) Mueller-Hinton agar plates prepared for the Kirby-Bauer test

Chemical Used	No. of Colonies from Spontaneous Mutations	No. of Induced Mutations	Determination of Whether the Chemical Is Highly, Moderately, Slightly, or Nonmutagenic
NEGATIVE CONTROL			

NAME _____ DATE _____ SECTION _____

QUESTIONS

1. A broth culture of his⁻ *S. typhimurium* is inoculated onto an agar plate that does not contain histidine. No other chemicals (possible mutagens) have been added. After incubation, several colonies are observed. Explain.

2. A broth culture of his⁻ *S. typhimurium* is inoculated onto an agar plate that does not contain histidine. A strongly mutagenic chemical is added, but no evidence of mutations shows up after incubation. Why? (*Hint:* Review step 1 of the Ames test procedure and consider at which stage during the life cycle of a microbe a mutation would take place.)

3. The strain of his⁻ *S. typhimurium* is unable to make anyone ill, for it requires a histidine-rich environment for growth. Yet, extreme care should be taken to prevent contamination and potential infection when performing this test. Why?

4. *Serratia marcescens* is placed on the surface of a Nutrient Agar plate so that a "lawn" of growth occurs with no isolation colonies. After exposure to UV light, however, distinct colonies are observed. Explain how this can occur.

MATCHING

a. autotroph

b. Ames test

c. UV light

d. Kirby-Bauer test

e. backmutation

f. zone of inhibition

g. spontaneous mutation

h. induced mutation

_____ results in an organism able to synthesize a needed substance that it was previously unable to produce by itself

_____ used to test bacterial sensitivity to antibiotics

_____ used to detect chemicals capable of causing mutations

_____ an organism that has lost its ability to produce a needed nutrient, structural chemical, or growth factor

_____ mutation resulting from the presence of an extraneous substance

_____ mutation that is not associated with any external stimulus

MULTIPLE CHOICE

1. A culture of bacterial cells is exposed to a measured amount of UV light. Some of the exposed cells become no longer able to synthesize a substance needed for survival. Which is the true statement?

   a. this is an example of an induced mutation
   b. these cells are now auxotrophic
   c. UV light acts as a mutagenic agent
   d. all of these statements are true

2. The microbe used in the Ames test to detect the presence of mutagenic chemicals is:

   a. *Salmonella typhimurium*
   b. *Serratia marcescens*
   c. *Staphylococcus aureus*
   d. *E. coli*

3. UV light is capable of damaging DNA and:

   a. proteins   b. carbohydrates   c. fatty acids   d. cytoplasm

4. One reason for the control plate in the Ames test is to determine:

   a. whether the strain of bacteria is truly his⁻
   b. the rate of spontaneous mutations
   c. whether the histidine overlay was prepared properly
   d. sterility of the paper disks

5. Humans lack the enzymes to synthesize vitamin C. Humans are:

   a. vitamin deficient   b. front-mutated   c. auxotrophic   d. mutagenic

WORKING DEFINITIONS AND TERMS

Auxotroph An organism that has lost the ability to produce a substance needed for survival. It must rely on the environment to provide this substance.

Backmutation A mutation that returns a gene rendered inactive due to mutation into a functioning one.

Induced mutation A mutation caused by an outside stimulus.

Lawn Solid growth of bacteria across the surface of an agar plate.

Mutagenic agent (mutagen) Any substance or physical entity capable of producing a mutation.

Spontaneous mutation A naturally occurring mutation not influenced by any external source.

UV light Light in the wavelength beyond the violet end of the visible light spectrum. This light is able to cause significant damage to DNA molecules, specifically at the point of two adjacent thymine nucleotides.

Epidemiology

Objectives

After completing this lab, you should be able to:	**2.** Describe how airborne microbes can be a source of infection.
1. State the importance of handwashing in controlling infections acquired in the hospital.	

In microbiology, *epidemiology* is the study of how a specific infectious agent survives and spreads through a group of susceptible individuals. Knowledge of epidemiology is important because awareness of how a specific microbe gets into and out of a person and how that microbe is *transmitted* between other individuals is the basis of developing methods of its control.

Two of the most common ways such an infectious agent can enter the body is through the respiratory and digestive systems—thus the rule of no eating or drinking in lab. Such an entranceway is called *a portal of entry.* Control over what enters the mouth and nose can significantly reduce the number of individuals who develop these respiratory and digestive infections.

Handwashing is perhaps the single most important method of controlling microbes in the clinical setting. When performed properly and consistently, the health care worker will transfer few, if any, microbes from patient to patient. This form of transferring microbes is called person-to-person or *direct contact. Fomites* are another source of microbial infection in the clinical area. A fomite is an inanimate (nonliving) object such as a fork or plate, thermometer, or urinary catheter, which is contaminated with an infectious agent. When handled improperly, it contaminates the health care worker who then often passes on the microbe to the patient. Proper and consistent handwashing also controls fomite transmission.

Respiratory secretions and the air itself (airborne transmission) are also major sources of infectious material. Properly utilized air filtration devices and masks are both effective in controlling the transmission of airborne microbes. Many infections are transmitted to the respiratory mucosa by the hands rather than directly by air. Microbes present on the fingers by direct or indirect (fomite) contact may be introduced into the eye, nose, and mouth by accidental touch. This is yet another reason for following strict aseptic technique procedures while handling infectious materials.

Very often personal care products can act as fomites or even reservoirs of infection. Although a *fomite* is a source of infectious microbes, it doesn't allow them to grow. A *reservoir of infection* allows the microbe to survive, grow, and maintain its ability to cause disease. Most personal care products such as nose spray and cosmetics have bacteriostatic agents added which impede the growth of contaminating bacteria. If no such agents were employed, a few contaminating microbes, well below any concentration needed for a successful infection, would soon increase to dangerous levels. Those who wish to see how many extra ingredients are included in their makeup and other materials that comes in contact with their bodies can easily test these products during this laboratory exercise.

EPIDEMIOLOGY

HANDWASHING PROCEDURE
(done with three students)

Materials List per Table/Work Station

Two Blood Agar Plates per student

Liquid disinfectant soap

Surgical scrub brush

PROCEDURE

1. Take two Blood Agar Plates and label the base of the plates as shown in Fig. 15.1. (This is done by each of the three students.)

2. Follow the handwashing procedure shown in Fig. 15.1.

3. Place your fingertips onto the agar section labeled "dry" with enough pressure to leave fingerprints. Try to avoid breaking or cracking the agar surface.

4. Rinse your hands (especially the fingertips) under running water while rubbing them together. After 30 seconds, shake off the excess water and press your fingertips onto the section of the plate labeled "wet" with enough pressure to leave fingerprints.

5. Wash your hands using one of the following three procedures for two more minutes and press your fingertips onto the appropriately labeled section of the second Blood Agar Plate. Then wash your hands for three more minutes and do the same. (Total of five minutes of handwashing.)

   a. One student will continue using plain water for the handwashing.

   b. One student will wash with the liquid disinfectant soap used in the laboratory. Rinse off the soap before pressing your fingertips onto the Blood Agar Plate.

   c. One student will use the hospital scrub brush saturated with disinfectant soap. Rinse off the soap before pressing your fingertips on the Blood Agar Plate.

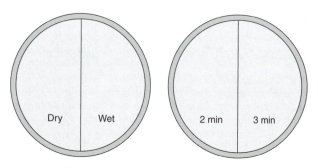

FIG. 15.1. Handwashing procedure.

Note: If available, a fourth plate can be prepared using regular soap.

6. When completed and properly labeled, place the Blood Agar Plates in the incubation tray.

Results: Handwashing

Observe the growth patterns on the Blood Agar Plates. Note the pattern of growth from the dry and wet fingertips. Compare the pattern of growth from the water-only washing to the disinfectant soap washing and the disinfectant soap-scrub brush washing. Observe not only the number of colonies seen but also the different types of colonies present.

Refer back to Figure 15.1 and sketch relative amounts of growth.

FOMITE AND DIRECT TRANSMISSION OF MICROBES

Materials List

Broth culture of *Micrococcus luteus*

T-Soy or Mueller-Hinton agar plate per student

Latex or vinyl gloves

One large test tube

PROCEDURE: CLASS

1. Each student will put on one glove and line up in one of two rows. *Important:* Make sure you use only the gloved hand for this part of the exercise. (See Fig. 15.2.)

2. One test tube contaminated on the exterior with *M. luteus* will be placed into the gloved hand of the student who will act as the "source of infection."

3. The "source of infection" will than pass the tube to the next student in the first row who will then pass it on to the next and so on.

4. The "source of infection" will then shake hands with the first student in the second row. The first student in the second row will then shake hands with the second student; the second with the third; the third with the fourth, and so on.

5. After each student's glove has been exposed to the contaminated tube, the glove will be pressed gently onto the surface of the assigned agar plate.

Note: Make sure the plate is properly labeled so that we can ascertain each person's method of becoming contaminated and his or her position in the row.

6. Once the above steps have been completed, each student should aseptically remove the glove and discard it in the autoclave bag. Follow the instructor's directions and wash your hands immediately afterward.

7. Once completed, the instructor will rinse off the tube with disinfectant and test for viable bacterial by inoculating a sample from the tube onto an agar plate. (See Fig. 15.2.)

Results: Fomite and Direct Transmission

The plates will be collected and placed in the same order as they were inoculated by the fomite (test tube) or direct transmission (handshake). Observe the pattern of growth on the plates from the first person to the last. (See Fig 15.3.)

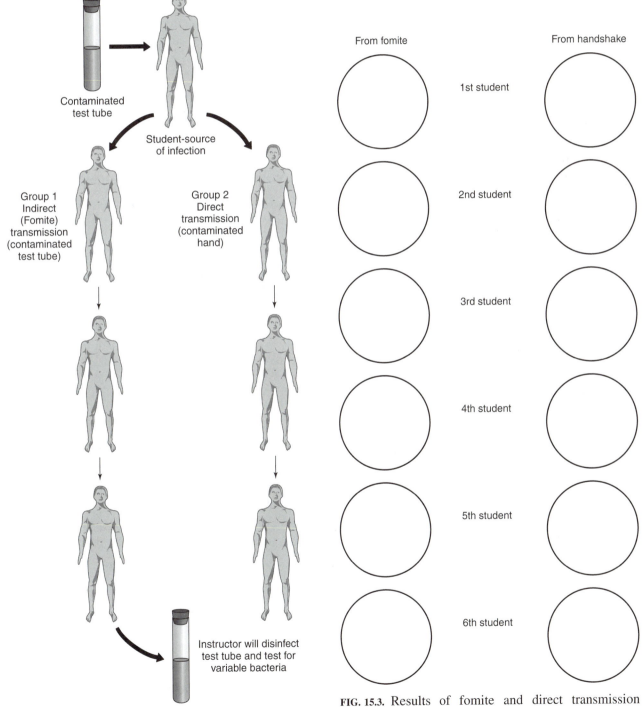

FIG. 15.2. Fomite and direct transmission.

FIG. 15.3. Results of fomite and direct transmission procedure.

AIRBORNE INFECTIONS: COUGH AND SNEEZE PLATES

PROCEDURE: AIRBORNE PLATES

1. Each table will take one Blood Agar Plate and expose it to the air. Allow it to stay open the rest of the laboratory period.

2. Record the amount of time the plate was left open.

    Time: _____

3. Label appropriately and place the plate in the incubation tray at the end of the period.

PROCEDURE: COUGH AND SNEEZE PLATES

Anyone in the class with a cough or sneeze due to a cold or allergy is volunteered for this procedure.

1. Label a Blood Agar Plate and keep it covered.

2. Every time you have to cough or sneeze, open the plate, hold it approximately six inches away from your mouth, and cough or sneeze into the plate.

3. Record how many times you coughed or sneezed on the plate.

    Number of Times _____

4. Label the plate appropriately and place it in the incubation tray at the end of the period.

Note: If this procedure is performed, one other person without a cold or an allergy will artificially cough or sneeze the same number of times on another Blood Agar Plate to act as a control.

Results: Airborne and Cough/Sneeze Plates

Observe the plates allowed to stay open during the period and note the amount and type of growth seen.

If cough or sneeze plates were performed, compare the amount of growth between the plates prepared by students with a genuine cough or sneeze with the controls. (See Fig. 15.4.)

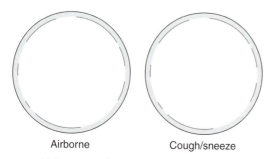

Airborne   Cough/sneeze

FIG. 15.4. Airborne and cough/sneeze plates.

MICROBES IN MAKEUP (OPTIONAL)

Materials List

One or two Blood Agar Plates per student

Test tube of sterile water or saline

Sterile swabs

Any cosmetics or personal care products provided by students

PROCEDURE

Select a personal care product such as a comb, brush, lipstick, mascara, eye-liner, or nose spray. Use a sterile swab for each product used. If the test material is dry, moisten the swab with sterile water or saline. Touch the swab to the item and then spread the sample onto the Blood Agar Plate. If you wish to test a liquid sample, place a drop on the Blood Agar Plate and streak through it using a sterile swab or an inoculating loop.

Results: Microbes in Makeup

Observe the presence of growth from any cosmetics or other personal care products on the Blood Agar Plates. Compare these results with others at your table and determine which products, if any, show evidence of harboring significant amounts of microbes.

Inventory

At the end of the laboratory, the following plates will be ready for incubation:

Two Blood Agar Plates for handwashing

One agar plate per student used for the fomite and direct transmission procedure

One Blood Agar Plate/table or work group for airborne microbes procedure

Cough or sneeze Blood Agar Plates and control plates as assigned

One or two Blood Agar Plates inoculated with personal care products (optional)

NAME _____ DATE _____ SECTION _____

QUESTIONS

1. After observing the growth of bacteria on the transmission plates, determine which mechanism allowed for the passage of more microbes: direct contact or fomites? Did the number of microbes decrease after a series of transfers between people occurred?

2. Explain why there is often more growth on the handwashing plates *after* the hands are washed than *before*.

3. Observe the growth of bacteria on the airborne and cough/sneeze plates. Can the air be a major source of infectious material in crowded conditions?

MATCHING

a. direct transmission

b. airborne transmission

c. reservoir of infection

d. portal of entry

e. fomite

f. epidemiology

_____ cough or sneeze

_____ a nonliving object capable of allowing a microbe to survive but not reproduce

_____ handshake

_____ allows an infectious microbe to grow, reproduce, and remain dangerous

_____ study of how an infectious agent spreads among a population

_____ method that a microbe utilizes to enter a susceptible host

MULTIPLE CHOICE

1. A partially filled coffee cup, including milk and sugar, is left standing overnight. It is contaminated with an infectious microbe. A person handles the *outside* of the cup and becomes contaminated. This is an example of a(n):

   a. fomite transmission   b. direct transmission   c. reservoir of infection   d. all of these

2. A partially filled coffee cup, including milk and sugar, is left standing overnight. It is contaminated with an infectious microbe. Some of the stale coffee-milk-sugar spills on a person's hand. This is an example of a(n):

   a. fomite transmission   b. direct transmission   c. reservoir of infection   d. all of these

3. The most important procedure for preventing the transmission of a microbe in the clinical area is:

   a. covering coughs and sneezes   b. handwashing

   c. use of air filtration   d. elimination of the portal of entry

4. A restaurant worker is polishing silverware with a towel just used to wipe down a dirty tabletop. The silverware is now a:

   a. vector   b. reservoir of infection   c. fomite   d. portal of entry

WORKING DEFINITIONS AND TERMS

Cross-contaminate To spread more than one microbe through handling or by fomites.

Epidemiology The study of the spread of diseases within a specific population or group.

Fomite A nonliving object capable of allowing a microbe to survive but not reproduce, that is, table, pen, test tube, fork.

Portal of entry Entranceway through which a specific microbe is able to invade a host.

Reservoir of infection Any object, living or nonliving, that allows a microbe to grow, reproduce, and maintain its ability to remain infective.

Vehicle of transmission An inanimate object or substance that allows an infectious agent to be delivered to individuals in a population, for example, air, food, and water.

Objectives

After completing this lab, you should be able to:

1. State the most practical methods of collecting and sending specimens to the microbiology laboratory.

2. List at least five errors that may occur in the collecting and sending of specimens to the microbiology laboratory.

3. Explain why transport medium is the preferred method of sending specimens to the microbiology laboratory.

4. Explain what a quantitative culture is and perform such a procedure on a real or simulated urine sample.

5. Categorize growth from a quantitative urine culture as no significant growth, infection, or acute infection.

6. Explain why a urine sample meant for bacterial examination should not be transported in a container used for biochemical testing.

7. Use the streak plate technique to isolate a Gram negative bacterium from a mixed culture.

16

EXERCISE

The proper collection and processing of laboratory specimens is an integral part of hospital procedures. If the laboratory receives an incorrectly collected and processed specimen for analysis, correct patient care may be jeopardized. A somewhat succinct computer programming term for something similar to this is: "Garbage in—garbage out." Virtually everyone working in the hospital or clinical setting will collect, handle, or transport some type of clinical specimen at one time or another. Therefore, it is imperative that an understanding of the procedures or protocols involved in these tasks be understood. **The best laboratory technologist in the world will not be able to accurately identify an infectious agent from a poorly acquired or transported specimen, and it is the patient who will suffer!**

The following principles apply to most microbiology laboratory specimens:

- The specimen must be collected before starting antimicrobial therapy. (*Streptococcus pyogenes* often does not show up in throat cultures as little as two hours after the start of antimicrobial therapy.)

- A specimen must be representative of the condition; for example, a sputum sample is more appropriate for testing for pneumonia, whereas saliva is not. Saliva (spit) is representative of the mouth, whereas sputum represents microbes from the lungs.

- The stage of the disease must be considered. For example, the typhoid fever bacillus is more easily isolated from blood during the first week of the disease and from the feces during the third and fourth week.

- Geography and season are considerations in determining which types of tests are to be performed. Plague is more common in New Mexico than in Alaska, and meningitis is more frequent in the winter than in the summer.

- Adequate amounts of material must be collected; otherwise not all tests can be performed.
- Proper transport is vital. If the specimen is allowed to dry out or to grow, lab results will be inaccurate.
- A series of samples may be necessary. If blood cultures are required, often a series of six to eight samples may be required, taken at two-hour intervals.
- The aseptic technique and sterile containers must be used for specimen collection and transport.
- Specimens MUST be kept moist or vegetative cells will die.
- Specimens MUST be labeled properly. Unlabeled or mislabeled samples are useless to the laboratory.

Hospital procedural manuals, as well as various malpractice trial records, are filled with numerous examples of how poorly collected and transported specimens contribute to the detriment of the patient. In this exercise, we will simulate correct and incorrect specimen collection procedures. You will be performing certain procedures involving handling and transporting specimens following accepted hospital techniques, as well as deliberately performing these techniques improperly to demonstrate the results of such errors.

Regardless of the type of specimen collected, a pure culture is usually necessary for accurate identification of most pathogens. Specimens sent to the laboratory tend to be mixed cultures of the organism causing the disease and the patient's normal or resident flora. The specimen sample will, therefore, be streaked on agar plates chosen to select and differentiate among the varying types of organisms associated with a particular body system or region. For example, a Blood Agar Plate will always be used with samples from the respiratory system, and MacConkey and/or Eosin Methylene Blue agar would be among the media used to isolate microbes from the gastrointestinal tract. Once colonies are isolated in pure cultures, the genus and species of the infective agent are readily identified.

THROAT CULTURES

One of the more common microbiological specimens taken in physician's offices and clinical settings is a throat culture. Because of the presence of pathogenic streptococci, a Blood Agar Plate is almost always used for such a sample. *Streptococcus pyogenes,* the causative (etiologic) agent of strep throat, is strongly beta-hemolytic and is readily identified by observing its growth on such a plate. A specimen is taken for a throat culture by touching a swab to the back of the patient's

throat. A tongue depressor is sometimes used to make sure the swab does not touch any area other than the target. Keep in mind that the specimen must be representative of the condition! Once taken, the swab may be processed in one of three ways:

1. *Direct culture.* The sample is placed directly on growth media, that is, a Blood Agar Plate. Since the time spent in transport is only two to three seconds, even the most delicate, or fastidious, microbe will survive the transition from throat to culture medium. This is the best way to prepare a culture, but it is often impractical. It is not convenient and too time consuming to send down to the lab each time a culture is needed, and media stored around various locations of the hospital often become outdated or contaminated.

2. *Transport medium.* This type of medium is often employed to alleviate the problem just mentioned. The various types of transport media prevent bacterial growth (bacteriostatic) and often contain chemicals such as charcoal, which absorbs bacterial toxins. Such a medium allows the lab to work on a representative sample of the microbes taken from the patient. If growth is allowed, the normal or indigenous flora (microbiota) may overgrow and inhibit the suspected pathogen. Toxins released from the normal or indigenous flora may also contribute to the inhibition of the pathogen. If placed in normal growth media, the pathogen's isolation and subsequent identification will often be difficult.

3. *Incorrectly processed.* Unfortunately, this aspect of specimen collection and transport is all too common. Of the dozens of incorrect procedures possible, allowing the sample to dry out is often a problem encountered by the laboratory receiving the cultures. This will be the incorrect procedure you will simulate as part of this laboratory.

Materials List per Table/Work Station

One tube of Stuart's transport medium
One dry, sterile test tube
Three Blood Agar Plates
Tongue depressors
Three sterile swabs

PROCEDURE

At the beginning of the laboratory period, one person from each table will be the "patient" and have two throat swabs taken by another person from the same table. The person taking the sample will carefully touch the back of the "patient's" throat with a sterile swab using a tongue depressor if necessary. (Remember: Just touch the back of the throat; you are getting a throat sample, not finding the gag reflex!) See Fig. 16.1 Care must be taken to avoid contaminating the swab by touching it to the tongue or tonsil. One swab will be placed in a tube of transport medium and the other in a dry, sterile test tube. Both tubes will be labeled and placed in the incubator until the end of the period. This will enhance the drying process previously mentioned.

At the end of the period, the "patient" will present his or her throat for a third throat swab. This last swab will be placed immediately on a Blood Agar Plate and streaked for isolated colonies (direct culture). The same process will be repeated with the swab preserved in transport medium and with the swab that has been allowed to dry out. When completed, each group will have three Blood Agar Plates of throat cultures streaked for isolated colonies.

STREAKING PROCEDURES

1. Place the swab with one throat sample on the Blood Agar Plate along the outer edge of the agar. *Rotate* the swab so that all surfaces of the swab come into contact with the agar. Once completed, discard the swab in disinfectant. (See Fig. 16.2.)

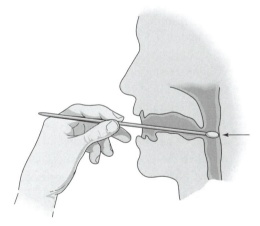

FIG. 16.1. Throat swab procedure.

FIG. 16.2. Rotate throat swab on the Blood Agar Plate, then use a loop to streak for isolated colonies.

2. Flame a loop, allow it to cool, and streak *through* the previously inoculated site *several times.*

3. Continue streaking the plate as you would a standard streak plate except you do not have to flame the loop between each section. In fact, you can go through the previously streaked area two or three times rather than just once. This is because of the relatively few microbes placed on the plate compared to inoculating bacteria from another plate or a slant.

4. Repeat the above steps with the remaining swabs on separate Blood Agar Plates.

5. Label and place the plates in the incubation tray.

Results: Throat Culture

Observe the three throat cultures: direct, transport medium and "dry." (This would also be a good time to review types of hemolysis.) The direct and transport medium cultures should be identical as to the *morphological types* of bacteria seen. In other words, if you observe five different colonies on one plate, you should see the same types of colonies on the other.

Note: You are observing the *type* of growth and not the *amount*. The amount of growth is a function of how the plate was streaked.

The plate inoculated from the swab that was allowed to dry out should not have as many colony types as the other two. This will be especially evident if it was possible to allow the plate to dry out for at least four hours.

QUANTITATIVE URINALYSIS

Urinary tract infections (UTI) are the most common type of infection found among hospital patients. The presence of significant amounts of bacteria in an individual's urine is indicative of such an infection. Most urine samples are collected through the natural process of urination. The urine flows through the urethra, which normally contains high numbers of certain microbes around the outer urethral orifice. Some of these organisms invariably turn up in the sample and must be considered whenever a urine sample is processed. Therefore, the *amount* of bacteria (quantity) is considered as well as the *type* of bacteria (quality). To eliminate this problem of extraneous bacteria in the urine as much as possible, certain collection techniques have been devised: the outer urethral area is washed with antiseptic soap, and the sample is collected in a sterile container via the *midstream* or *clean catch* technique. With this procedure, the patient begins to urinate, causing the initial flow of urine to flush out most of the normal flora around the urethral orifice. The sterile container is then placed in the flow of the remaining urine. Most bacteria found in the samples collected in this manner are indicative of microbes from the urinary bladder rather than the urethra.

Since there will always be some urethral flora in a urine sample, counting methods must be utilized to take these microbes into account. Before a diagnosis can be made, a basic method for performing such a quantitative culture is utilized. Several variations may be used to achieve such a culture. One involves the use of a *calibrated loop, smear plates,* and *glass spreading rods* (*"hockey sticks"*). Another procedure also uses a *calibrated loop* and a modification of the streak plate.

Samples of urine are usually inoculated on a Blood Agar Plate and a MacConkey or Eosin-Methylene Blue agar plate (MAC/EMB), smeared, or specially streaked over these media and allowed to grow. The calibrated loop used holds approximately 0.001 ml of urine. The quantity of bacteria found in 1 ml of urine can, therefore, be calculated by counting the number of colonies growing from the 0.001 ml sample and multiplying that number by 1000. Since a regular streak plate allows the bacteria to grow into each other in the first section of the plate, methods must be employed to avoid this phenomenon. Among these methods are a modified streak plate and a smear plate procedure. (The smear plate procedure is similar to that used to spread bacteria over the Mueller-Hinton plate in the Kirby-Bauer antibiotic sensitivity test, except that the bent glass "hockey stick" does not absorb bacteria as a sterile swab would.

The Blood Agar and MacConkey/Eosin-Methylene Blue (MAC/EMB) combination allows technicians to diagnose the Gram stain reaction of the bacteria without having to get their fingers stained. The determination is made as follows: (Remember: Everything grows on Blood Agar and only Gram negatives will grow on the others.)

- Growth on blood and no growth on MAC/EMB = Gram positive
- Equal growth on both types of plates = Gram negative
- Large amount of growth on blood and less growth on the other plate = both types

Very often, the growth seen is divided into several different categories based on quantity. The presence of *0–9 colonies* on the plate is usually reported as *no significant growth,* for such a small quantity indicates that the microbes came from the urethra. The presence of *10–99 colonies* indicates an *infection,* and *over 100 colonies*

TABLE 16.1	QUANTITATIVE EVALUATION OF URINE SPECIMENS		
	Colonies/ 0.001 ml	*Microbes/ ml Urine*	*Significance*
	0–9	0–9000	No significant growth
	10–99	10,000–99,000	Infection
	100+	100,000+	Acute infection

means an *acute infection* is present. If the bacterial growth completely covers the plate so that no individual colonies are discernible, the results may be read as: Too Numerous To Count (TNTC). (See Table 16.1.)

Since almost all urinary tract infections are caused by a single species (usually *E. coli*), the presence of several different microbes often indicates contamination or an improper collection technique. If personal urines are collected in this class, few bacterial colonies will be seen on the quantitative plates. By inoculating either your personal urine or the simulated urine with one loopful of a bacterial broth culture, an "infection" or "acute infection" growth pattern can be observed.

Urine samples are also sometimes mishandled. Besides the problems previously mentioned in the collection process, the sample may be allowed to stand too long before culturing, or it may be collected in an improper container. If allowed to stand at room temperature too long, the bacterial count will go up as growth naturally occurs, resulting in an inaccurate laboratory diagnosis. For this reason, urine samples are sent to the lab within two hours, or refrigerated, or collected in a container with a bacteriostatic agent.

If the wrong chemical is used in the container, the bacteria will be killed rather than preserved, once again resulting in a misdiagnosis. Containers used for biochemical tests on urine often contain disinfectants such as thymol or other bacteriocidal substances so that the bacteria will not change the chemical composition of the sample. For example, bacterial growth in a diabetic's urine sample may reduce the amount of glucose present (remember, phenol red broths, O.F. basal medium, and Triple Sugar Iron agar), as well as change the pH.

Materials List per Table/Work Station

Broth culture of *Serratia marcescens*
Bottle of thymol crystals

Materials/Person

One urine sample, either personal or simulated
One Blood Agar Plate
Three MacConkey or Eosin-Methylene Blue agar plates

One glass spreading rod ("hockey stick")

One calibrated loop

One sterile test tube

PROCEDURE

1. Acquire a urine sample, either your own or a simulated one provided by your instructor. If it is your own, you do not have to conform to the collection techniques previously mentioned. That is, you do not have to clean the outer urethra or perform a clean catch.

2a. Smear plate method for application of urine to a plate: Mix well if allowed to stand more than 30 minutes and use the calibrated loop to place 0.001 ml of urine in the center of the Blood Agar and MAC/EMB agar plates. (See Fig. 16.3.)

Smear the urine sample on each type of agar plate with the glass hockey stick as demonstrated. See Fig. 16.4 (This initial sample will probably show up as "no significant growth" even if a sterile container is not used, the urethra is not cleaned, and a clean catch is not made.) Flame the glass rod for a few seconds and return it to the test tube for reuse or dip in an alcohol solution for several seconds.

⚠ **CAUTION:** OPEN CONTAINERS OF ALCOHOL SHOULD NOT BE NEAR OPEN FLAMES.

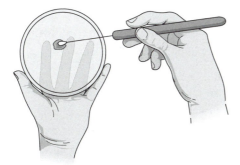

FIG. 16.3. Place urine sample on agar plate using a calibrated loop.

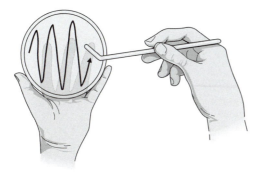

FIG. 16.4. Smear urine sample over agar plate using a sterile, glass spreading rod ("hockey stick").

2b. Alternative method for application of urine to a plate: Mix well if allowed to stand more than 30 minutes and use the calibrated loop to inoculate 0.001 ml of urine on the Blood Agar Plate and 0.001 ml of urine on the MAC/EMB plate. Make a single streak from top to bottom on each of the plates. (See Fig 16.5.) Then, using a regular loop, streak back and forth across the original streak until the measured urine sample is spread over the entire plate. (See Fig. 16.6.)

Note: Make sure you use a regular loop for the streaking: After step 2 is completed:

3. Pour some of the urine sample from step 2 into a sterile test tube. Purposely contaminate your sample with one loopful of *S. marcescens* broth. Mix well and prepare a second MAC/EMB agar plate, as described in step 2 above.

4. Simulate an incorrect procedure for processing the urine sample as follows:

   a. Add several crystals of thymol to the contaminated urine sample in the test tube used for step 2.

   b. Mix gently and allow it to sit for at least 30 minutes.

   c. Prepare a third MAC/EMB agar plate, as described in step 2 above. This sample will simulate collecting the urine in an inappropriate container with the incorrect additives, for example, bacteriocidal rather than bacteriostatic. The number of colonies grown on this plate should be considerably smaller than the one prepared in step 3 since thymol destroys bacteria.

5. Place the labeled plates in the incubation tray.

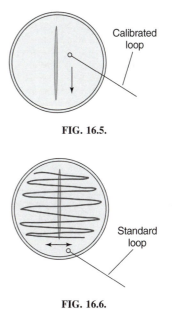

FIG. 16.5.

FIG. 16.6.

Results: Urine Samples

Observe the Blood Agar Plate and the MAC/EMB inoculated with the fresh urine sample (step 2). If you used your own urine, even without the aseptic technique procedures mentioned previously, there will probably be little, if any, bacterial growth (*no significant growth,* or 0–9 colonies).

The second MAC/EMB plate prepared with the contaminated urine sample should show evidence of an *infection* (10–99 colonies) or an *acute infection* (100 + colonies). If there is so much growth that individual colonies are difficult to observe, the results may be reported as Too Numerous To Count (TNTC).

Finally, observe the third MAC/EMB plate for growth. There should be less growth seen here than on the second MAC/EMB plate, for this was the simulation of improper transport technique.

SPECIMENS FROM THE GASTROINTESTINAL TRACT

Since the lower GI tract contains the highest concentration and numbers of different bacteria in the body (fecal material is over 50% bacteria and may contain over 300 different species), samples taken from this region will always contain large numbers of microbes other than the one being tested for. The medical microbiology laboratory must be able to isolate such bacteria from fecal samples by using appropriate selective and differential media. You will practice such an isolation procedure by streaking a mixed bacterial sample (simulated intestinal bacteria) on the appropriate media. Once isolated, you will attempt to identify the genus and species of one or perhaps both of the isolated microbes using techniques covered in future exercises.

Materials/Person

One simulated GI tract specimen in transport medium

One plate each of T-Soy, MacConkey, and Eosin-Methylene Blue agar

PROCEDURE

1. Inoculate the sample on to all three isolation plates in the same manner as described for the throat cultures.

2. Streak for isolation. Make sure you remember the unknown number.

3. Label the plates with your name and unknown number and place in the incubation tray.

Results: GI Tract Isolation Technique

Inspect the three plates you used for isolation. Look for isolated colonies. Determine whether:

- Any of the microbes are able to swarm (T-Soy plate).
- The MacConkey agar prevented the growth of any of the microbes. (MacConkey does inhibit some Gram negatives.)
- There is green metallic growth on the Eosin-Methylene Blue agar.

Choose a plate that shows well-isolated colonies. You will utilize these isolated colonies to inoculate media used to identify enteric pathogens. Pick one of the colonies and inoculate a fresh T-Soy agar plate. Streak for isolation. Follow your laboratory instructor's direction as to subculturing. If only one microbe is to be further identified, subculture that one onto a separate T-Soy plate. If both microbes are to be identified, inoculate two separate plates with the two different colonies.

Inventory

At the end of the laboratory, the following will be placed in the incubation tray:

Each group
Three throat culture Blood Agar Plates
    One direct culture
    One from transport medium
    One allowed to dry out
Each person
    Four quantitative urinalysis plates (one Blood Agar Plate, three MAC/EMB plates)
    Three GI tract isolation plates (T-Soy, Eosin-Methylene Blue, MacConkey)

NAME _____ DATE _____ SECTION _____

QUESTIONS

1. What are transport media and why are they useful?

2. What are the dangers of allowing bacteria to grow before they can be processed by the laboratory?

3. List 10 improper techniques or procedures associated with the collection and transport of microbiological specimens.

4. How can one determine the Gram stain reaction of a bacterium without performing the stain itself?

5. Why is the amount of bacteria in a urine sample as important as the type of bacteria?

MATCHING

a. acute urinary tract infection

b. 0.001 ml

c. 9000 bacteria/ml

d. diagnosis of Gram positive infection

e. direct culture

f. transport medium

g. diagnosis of Gram negative infection

h. midstream or clean catch

i. 0.01ml

j. 25,000 bacteria/ml

_____ preserves microbiological specimen until it reaches the laboratory

_____ considered "no significant growth" in a quantitative urinalysis

_____ placement of a microbiological specimen from the source, immediately onto growth medium

_____ urinalysis plates showing the same amount of growth on a Blood Agar Plate and on an Eosin-Methylene Blue plate

_____ quantitative plate count of greater than 100 colonies from a calibrated loop

_____ method used to eliminate much of a patient's normal flora from a urine sample

_____ urinalysis plates showing growth on a Blood Agar Plate and no growth on a MacConkey agar plate

_____ amount of urine delivered to an agar plate by a calibrated loop

MULTIPLE CHOICE

1. A urine sample shows high concentrations of bacteria, but the individual tested shows no symptoms. All subsequent samples show no growth. A possible error for this first test was:

   a. the sample was taken incorrectly
   b. the sample was not processed quickly and not refrigerated
   c. the sample was taken from the wrong patient
   d. all of these

2. Transport medium does all of the following except:

   a. inhibits nonpathogens and allows pathogens to grow
   b. is bacteriostatic
   c. absorbs certain toxins
   d. keeps pathogenic microbes alive until they can be processed

3. Which of the following should be performed or utilized in collecting a urine sample for bacterial evaluation?

   a. use of a sterile container
   b. cleansing of the outer urethra
   c. clean catch or midstream catch technique
   d. all of these

4. A GI tract sample is collected and delivered to the lab. The medium used to isolate a suspected Gram negative etiologic agent is:

   a. MacConkey agar
   b. Phenyl Ethyl Alcohol agar
   c. Triple Sugar Iron agar
   d. Sulfide Indole Motility Medium

5. A Gram positive microbe is suspected of causing a urinary tract infection. Growth on which of the following will confirm this suspicion?

   a. Blood Agar Plate   b. MacConkey agar   c. Eosin Methylene Blue   d. all of these

6. Which of the following will absorb bacterial toxins?

   a. Blood Agar Plate   b. transport medium   c. Eosin-Methylene Blue agar   d. MacConkey agar

7. The most practical way to deliver microbes to the laboratory is:

   a. T-Soy broth   b. transport medium   c. thiolglycollate   d. dry, sterile test tube

8. A quantitative urinalysis sample shows over 200 colonies on an Eosin-Methylene Blue plate. Most of these colonies grow with a green metallic color. (Review previous labs for a hint.) Which of the following is most likely?

   a. a chronic infection by a Gram positive
   b. an acute infection by a Gram positive
   c. an acute infection caused by *E. coli*
   d. an infection caused by *Proteus mirabilis*

WORKING DEFINITIONS AND TERMS

Calibrated loop Inoculator whose loop size is adjusted to consistently hold a specific volume of liquid.

Clean catch Technique of collecting a urine sample, which eliminates much of the outer normal urethral flora. The procedure includes cleaning the outer urethral region and acquiring a midstream sample.

Midstream collection Collecting a urine sample while in the process of urination. See "clean catch."

Too Numerous To Count (TNTC) Term used in quantitative cultures to indicate that an excessively large number of microbes are growing on an agar plate, thus making it impossible to estimate their number.

Quantitative culture Culture technique that allows you to determine the number of microbes in a measured sample as well as their type.

Transport medium Bacteriostatic medium used to transfer bacterial specimens to the laboratory. It prevents overgrowth of the pathogen by normal flora and may contain toxin-absorbing compounds.

17 Specific Laboratory Tests

Objectives

After completing this lab, you should be able to:

1. Name tests that will specifically identify *Staphylococcus aureus, Staphylococcus saprophyticus, Streptococcus pyogenes,* and *Streptococcus agalactiae.*

2. Identify the test that will differentiate between streptococci and staphylococci.

3. Identify the test that will differentiate Gram negative rods between the family *Enterobacteriaceae* and most *Non-Enterobacteriaceae.*

4. Define the term *presumptive identification.*

Single-test, rapid identification techniques for specific pathogens are an integral part of the repertoire of procedures performed by laboratory technicians. If the physician suspects a specific microbe as the cause of a disease process, the lab may be requested to test for that microbe along with all the other standard tests. Other tests can rapidly categorize microbes into specific groups or genera. Serological tests (use of antibodies and antigens, e.g., blood typing) make up an important part of these rapid identification methods and will be covered in another exercise.

Is it "staph" or "strep"? In the earlier part of this course, you may have struggled with identifying the morphological arrangement of these genera under the microscope. If you took a sample of streptococci and mixed it too thoroughly before staining, it often looked like staphylococci under the microscope. The *catalase* test will rapidly differentiate between these two genera by the use of hydrogen peroxide (H_2O_2).

Catalase is necessary in most cells with an aerobic metabolism because one of the byproducts of oxygen metabolism is formation of hydrogen peroxide, which is extremely toxic. Catalase will rapidly catabolize hydrogen peroxide into oxygen and water. When done on

a larger scale outside the cell, bubbling will occur. Since red blood cells contain high concentrations of this enzyme, this explains the bubbling when hydrogen peroxide is poured on a wound. The aerotolerant streptococci are negative for catalase (one reason for their small colonies) and the staphylococci are positive. A few drops of hydrogen peroxide can easily allow one to distinguish between these two genera.

The *oxidase test* can rapidly differentiate between two major groups of Gram negative rods: *Enterobacteriaceae* and *Non-Enterobacteriaceae. Escherichia coli* is a typical example of the first group, and *Pseudomonas aeruginosa* belongs in the second. Since all members of the *Enterobacteriaceae* are oxidase-negative and most members of the *Non-Enterobacteriacea* have this enzyme, an appropriate test will quickly distinguish between these two groups.

The *coagulase* test specifically identifies "pathogenic" *Staph aureus.* Coagulase is a virulence factor that causes blood to clot or coagulate under laboratory conditions. Although many feel that coagulase does not react in this manner in the host, the enzyme has been shown to interfere with phagocytosis, thus protecting any microbe that secretes it. Many laboratories divide the staphylococci into only two groups: coagulase positive and coagulase negative. The coagulase positive organisms are considered to be the pathogen *S. aureus,* and the other several species are grouped together as nonpathogenic and are not usually considered as part of the clinical problem.

There is another test used to **presumptively** identify a specific species of staphylococci other than *S. aureus.* Coagulase negative *S. saprophyticus* is a common urinary tract infectious agent in sexually active females. While *S. aureus* is capable of causing disease symptoms in nearly all regions of the body, *S. saprophyticus* infections are usually confined to the urinary tract. If a quantitative urinalysis shows a high concentration of suspected staphylococci, this species must be suspected. One can readily differentiate between *S. saprophyticus* and most other clinically important staphylococci by measuring resistance to the antibiotic *novobiocin.*

> "Presumptive" identification indicates that the test is highly but not 100% accurate for a particular microbe. Any presumptive identification should be confirmed by additional tests. For example, metallic green growth on EMB is presumptive of *E. coli* since this microbe usually displays this type of growth on EMB. However, other microbes, such as *Klebsiella pneumoniae,* sometimes grow with this same metallic green sheen.

Bacitracin sensitivity has long been a traditional test to presumptively identify *Streptococcus pyogenes,* which is also known as Group A, beta-hemolytic strep. Of all the bacteria capable of causing acute pharyngitis (sore throat), only one is β-hemolytic and extremely sensitive to bacitracin. By growing the suspected microbe on a Blood Agar Plate and placing a bacitracin disk in the middle of the inoculation, *S. pyogenes* can be presumptively identified. The bacitracin disk is not the type used in the Kirby-Bauer technique of antibiotic sensitivity covered in an earlier lab. The disk used for this test contains a minimal amount of the drug. Even this tiny amount of bacitracin will inhibit enough of the bacteria to allow a red zone of undamaged red blood cells to exist within the completely clear (β-hemolytic) area of growth. (See Fig. 17.1.)

The *CAMP Test* (named after its developers Cristie, Atkins, Munch, and Peterson) is a presumptive test for Group B strep or *S. agalactiae.* This microbe is rarely dangerous to a healthy adult but is a known pathogen for newborns. The principle of the test is based on the moderate β-hemolytic ability of certain strains of *S. aureus* and the extremely weak β-hemolytic activity of Group B strep. *S. aureus* releases its hemolytic enzymes such that the area immediately adjacent to its growth is

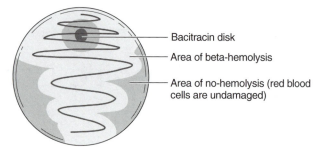

FIG. 17.1. Presumptive test for *Streptococcus pyogenes.* Note the lack of β-hemolysis surrounding the bacitracin disk.

completely clear while the outer regions still contain some unhemolyzed cells. The enzymes of Group B strep enhance the action of the staphylococcus. When these two groups of enzymes combine, the red blood cells in the region will be completely hemolyzed. If both microbes are placed so that the zones of hemolysis overlap, the arrowhead or mushroom cap-shaped clear area of β-hemolysis is indicative of *S. agalactiae.* This is known as a synergistic effect where the combination of these two enzymes working together cause a reaction much greater than either one by itself. (See Fig. 17.2.)

Materials/Class for Demonstration

Agar plate culture of *Streptococcus pyogenes*

Bacitracin disks

Blood Agar Plate

Materials List per Table/Work Station

Agar plate cultures of *Staphylococcus aureus, Staphylococcus saprophyticus , Staphylococcus epidermidis, Streptococcus agalactiae, Enterococcus faecalis, Escherichia coli, Pseudomonas aeruginosa*

Blood Agar Plates

Small test tubes

Novobiocin disks

Wax film for covering small test tubes

3% hydrogen peroxide

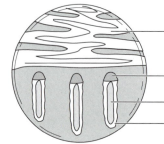

FIG. 17.2. The CAMP Test.

Oxidase test reagent

Rabbit coagulase plasma

Millimeter ruler

BACITRACIN SENSITIVITY— DEMONSTRATION

Note: There are no attenuated or avirulant strains of *Streptococcus pyogenes* available. If not done as a demonstration, extreme care must be utilized.

Samples of *S. pyogenes*, *S. agalactiae*, and *E. faecalis* are streaked on a Blood Agar Plate so that there will be several areas of solid bacterial growth with no isolated colonies. Make sure you keep the inoculations well separated. A bacitracin disk will then be placed within each zone of growth. *S. pyogenes* is β-hemolytic, and since it is **extremely** sensitive to bacitracin, a zone of inhibition around the bacitracin disk will show up red within the clear zone of hemolysis. This phenomenon is highly indicative of this microbe. (See Fig. 17.1.)

Results: Bacitracin Sensitivity

Observe the growth pattern of *S. pyogenes* on the Blood Agar Plate. Notice the zone of inhibition surrounding the bacitracin disk. Did any of the other streptococci tested show strong β-hemolysis as well as high sensitivity to bacitracin?

OXIDASE TEST

Take a piece of paper towel or filter paper and place two separate drops of oxidase test reagent (tetramethyl-para-phenylenediamine dihydrochloride) on it. Aseptically remove a sample of *E. coli* from the agar plate using the back of a cotton swab (either plastic or wood). Rub the sample on one of the drops of test reagent. (Using a metal loop may result in a false positive reaction.) Perform the same procedure with a sample of *Pseudomonas aeruginosa*. A purple color, which must develop within 30 seconds, indicates the presence of oxidase enzymes. Any color change after this time is to be considered negative.

Results: Oxidase Test

Determine which of the two Gram negative microbes is oxidase positive (*Non-Enterobacteriaceae*) and which one is oxidase negative (*Enterobacteriaceae*).

COAGULASE TEST

Two modifications of this test can easily be performed.

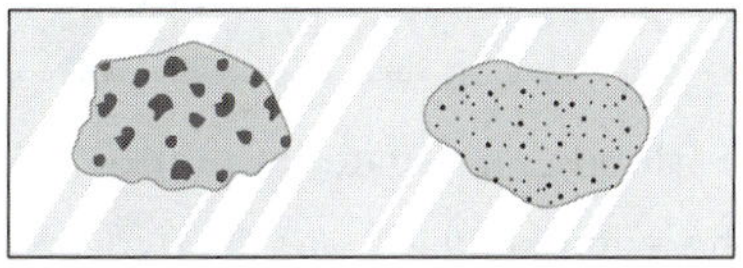

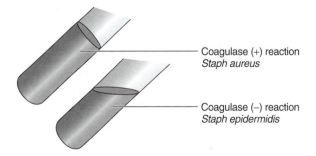

S. aureus   S. epidermidis

FIG. 17.3. Rapid slide test for coagulase.

Rapid Slide Test

1. Place two equal volumes of coagulase test plasma on a glass slide.

2. Aseptically place a sample of *S. epidermidis* in one of the volumes and mix well as you would for a stain preparation. You should notice that the mixture becomes uniformly cloudy with little or no clumping.

3. Repeat this procedure with *S. aureus* on the second plasma sample. Look for an obvious clumping of the bacteria indicative of a positive reaction. This is because the enzyme is often attached to the surface of the cell wall and its action causes the plasma to coagulate, resulting in large amounts of bacteria trapped together. (See Fig. 17.3.)

Tube Test

The tube test is considered more accurate than the rapid slide test because it will detect the enzyme even if it is not attached to the external cell wall. The disadvantage of the test is that it takes several hours of incubation to determine a reaction.

1. Aseptically place 2–3 ml of sterile coagulase plasma in each of two test tubes.

2. Inoculate a sample of *S. epidermidis* in one tube and *S. aureus* in the other.

3. Seal the tubes with tape or wax film and incubate.

Results: Coagulase Tube Test

Observe the two tubes inoculated with the staphylococci. Gently tilt them and determine whether any of the samples have solidified. If the sample is solid, the microbe inoculated is coagulase positive; if liquid, it is coagulase negative. (See Fig. 17.4.)

Coagulase (+) reaction
Staph aureus

Coagulase (–) reaction
Staph epidermidis

FIG. 17.4. Tube test for coagulase.

NOVOBIOCIN SENSITIVITY

1. Divide a Blood Agar Plate into thirds.

2. Touch the equivalent of one to two colonies of *S. saprophyticus* with a sterile swab and mix well in a tube of sterile saline.

3. Press the swab against the wall of the test tube and twist to remove excess moisture; cover one-third of the Blood Agar Plate with this sample. Take care not to overlap into the other two-thirds of the plate. (See the Kirby-Bauer procedure covered earlier for a review of this procedure.)

4. Repeat steps 2 and 3 with *S. aureus* and *S. epidermidis* so that all the bacteria are inoculated onto the same Blood Agar Plate.

5. Aseptically place a 5 μg novobiocin disk in the center of each of these inoculations; then touch each one with a sterile loop to ensure good contact. Invert the plate when completed.

6. Place in the incubation tray.

Results: Novobiocin Sensitivity

Observe the zones of inhibition surrounding each of the novobiocin disks. Any zone of inhibition equal to or less than 12 mm is considered resistant and therefore presumptive for *S. saprophyticus*. Any zone greater than 12 mm diameter indicates sensitivity and is therefore presumptively negative.

CAMP TEST

1. Take a sample of *S. agalactiae* and streak over one-half of a Blood Agar Plate so there is solid bacterial growth.

2. Streak three lines of *S. aureus* at right angles to the *S. agalactiae*. Make sure you do not overlap the inoculations. (See Fig. 17.2.)

3. Label and place in the incubation tray.

4. Repeat this procedure using *E. faecalis* as the test organism.

Results: CAMP

Observe the Blood Agar Plate inoculated with the *S. agalactiae* and *S. aureus*. Look for a region between the growth that shows an extra clear zone of β-hemolysis. Compare these results with the Blood Agar Plate prepared with *E. faecalis*. (See Fig. 17.2.)

CATALASE TEST (to be done when all other inoculations are completed)

1. Open the covers of all the samples of bacteria used in this lab.

2. Separate the plates containing streptococci from those containing staphylococci.

3. Add 1–2 drops of hydrogen peroxide to each of the cultures growing on the plates and look for the characteristic bubbling that indicates the presence of catalase.

An alternative to this method is to place 1 drop of hydrogen peroxide onto a slide, mix in a sample of the suspected bacterium, as you would a bacterial smear, and look for bubbling.

Note: It is not recommended that the catalase test be performed directly on any bacteria growing on a Blood Agar Plate. The catalase contained in the red blood cells will react with the hydrogen peroxide and can make an accurate reading difficult.

Results: Catalase Test

Determine which genus of the Gram positive bacteria was catalase positive and which one was catalase negative. Also determine whether any of the Gram negative bacteria contained this enzyme. Fill in the following chart.

Microbe	Reaction +/−
Staphylococcus aureus	
Staphylococcus saprophyticus	
Staphylococcus epidermidis	
Streptococcus agalactiae	
Enterococcus faecalis	
Escherichia coli	
Pseudomonas aeruginosa	

NAME _____ DATE _____ SECTION _____

QUESTIONS

1. How can you determine whether a staphylococcal culture is considered pathogenic?

2. How can you rapidly distinguish between a staphylococcus and a streptococcus?

3. Why does hydrogen peroxide bubble when it is placed on a wound or on certain types of bacteria?

4. Some microbiologists place a bacitracin disk in the center of the suspected *S. agalactiae* inoculation as part of the CAMP test. Why would this be done?

5. When can an antibiotic be used to help identify a specific microbe?

MATCHING

a. novobiocin

b. CAMP

c. hydrogen peroxide

d. bacitracin

e. oxidase

f. *Staphylococcus aureus*

g. staphylococci

h. streptococci

i. coagulase

_____ used to presumptively identify *Streptococcus pyogenes*

_____ used to presumptively identify *Enterobacteriaceae*

_____ contains the enzyme coagulase in its cell wall

_____ used in the presumptive identification of *Staphylococcus saprophyticus*

_____ genus of bacteria that is catalase negative

_____ microbe that is positive for a test that shows clumping of the microbe on a slide

_____ test used to identify *Streptococcus agalactiae*

_____ genus of bacteria that is catalase positive

_____ used to test for the presence of catalase

MULTIPLE CHOICE

1. Catalase reacts with:

   a. hydrogen peroxide   b. cell walls

   c. beta-hemolytic bacteria   d. alpha-hemolytic bacteria

2. An oxidase positive microbe would most likely be a(n):

   a. *Enterobacteriaceae*   b. *Non-Enterobacteriaceae*   c. staphylococcus   d. streptococcus

3. A microbe containing this enzyme is almost always considered a pathogen

   a. catalase   b. coagulase   c. oxidase   d. beta-hemolysis

4. Which of the following is strongly beta-hemolytic and extremely sensitive to bacitracin?

   a. *Staphylococcus aureus*
   b. *Staphylococcus epidermidis*
   c. *Streptococcus saprophyticus*
   d. none of these

5. Which of the following shows a characteristic resistance to novobiocin?

   a. *Staphylococcus aureus*
   b. *Staphylococcus epidermidis*
   c. *Staphylococcus saprophyticus*
   d. *Streptococcus agalactiae*

6. The CAMP test uses which of the following to identify *Streptococcus agalactiae?*

   a. hydrogen peroxide   b. *Staphylococcus aureus*   c. novobiocin   d. bacitracin

7. A Gram positive coccus is isolated in the lab. Which of the following tests would give us an idea as to its genus?

   a. coagulase   b. novobiocin   c. catalase   d. oxidase

8. A Gram negative rod is isolated in the lab. Which of the following tests would give us an idea as to the family?

   a. coagulase   b. novobiocin   c. catalase   d. oxidase

9. A suspected staphylococcus is isolated in the lab. Which of the following tests would be used to identify it as *S. aureus?*

   a. catalase   b. coagulase   c. bacitracin   d. novobiocin

10. An infant is suffering from suspected meningitis. Which of the following tests would be performed if *Streptococcus agalactiae* was suspected?

    a. oxidase   b. catalase   c. CAMP   d. novobiocin

18 Serology

Objectives

After completing this lab, you should be able to:

1. Explain the concept of specificity of antigen-antibody reactions.

2. Describe two different procedures, one of which can demonstrate the presence of antibodies in blood, and another which can identify an etiologic agent.

3. Distinguish between the terms *plasma* and *serum*.

4. Explain the significance of the presence of a particular type of antibody in an individual's blood.

5. Define agglutination.

6. Explain how an antigen can be detected by using known antibodies.

7. Explain the principle of the "ELISA" procedure.

Serology makes use of blood serum (plasma without the clotting factors). This serum may contain antibodies against disease-producing agents to which the individual has been exposed or against antigens used for vaccination. Antibody is produced only in response to the presence of a specific antigen (usually a microbe) in the body. Serological testing simply permits antibodies (from serum) and known antigens that are added to it to interact and for that interaction to be visualized in some way. This demonstrates the presence of the antibody, which would only be there because the individual was exposed to the antigen (disease agent). Antibodies directed against a certain *etiologic agent* almost always react with, and only with, that one agent. Thus, antibody-antigen reactions are described as having a high degree of *specificity*.

It is easy to reverse the procedure by taking known antibodies to react with and identify unknown microbes. Because of the specificity of antibody-antigen reactions,

such a procedure can identify the unknown organism. This provides rapid diagnostic information to the physician, who can then initiate appropriate antibiotic therapy.

So much knowledge has been accumulating about how antibodies form and the roles played by various cells (especially leukocytes) and biologic mediator substances that serology is no longer prominent in its own right, but has been absorbed and been made part of the modern science of immunology.

You will be made aware of three of the many techniques employed in immunological testing. One procedure tests serum for the presence of antibody against a known disease agent (antigen). The other two seek to identify an unknown microbe by seeing if it will interact with a known antibody. These tests are selected to enable the student to understand very basic aspects of serology. More sophisticated and more elegant methods than the ones described here are available in the clinical laboratory.

FEBRILE AGGLUTININS TEST: THE *BRUCELLA* SCREENING PROCEDURE

Febrile antigens generally refer to microbes, which cause fever in the host. *Brucella* species are examples of microbes that possess such a febrile antigen. Since the early 1900s, febrile antigens have been demonstrated by *agglutination* tests. Such tests make use of the "clumping together" of microscopic particles until they are visible to the naked eye. The visible agglutination seen is due to specific antibodies (especially multivalent IgM and divalent IgG) interacting with several different epitopes of the antigen. (See Fig. 18.1.) Blood typing also makes use of an agglutination technique.

Brucellosis (undulant fever) is caused by a Gram negative bacillus of the genus *Brucella*. The bacteria are endemic to farm animals, including cows, pigs, and goats. Historically, its incidence in humans is limited to a small number of farmers, veterinarians, slaughterhouse employees, and dairy workers. Cultures from infected individuals are often negative, so serologic testing can be helpful in early diagnosis of the disease. Antibiotic therapy is quite successful as a treatment. Symptoms include fever, arthralgia (painful joints), malaise, chills, and sweating. The most frequent complication in humans is osteomyelitis.

PROCEDURE

In this modified rapid slide test method, *Brucella* antigenic suspension is used to detect antibodies. Serum from the patient having the antibody will *agglutinate* the antigens. The presence of a significant titer (concentration) of antibodies to this antigen indicates at least exposure to this microbe.

⚠ **REMEMBER: DO NOT PIPETTE BY MOUTH.**

The *Brucella* antigen used in this exercise is a chemically inactivated, stabilized antigen suspension. The positive control and some of the unknown sera (patient sera) contain antibodies obtained from laboratory animals. A preservative is added to these substances to prevent contamination and extend their shelf life.

Materials List per Table/Work Station

Bacto® *Brucella* antigens and controls

Each pair of students will be supplied with three reagents. Some of the "patient serum" solutions will be positive, and some will be negative. **Record the identification code of the solutions assigned to you.**

1. Bring the antigen suspension, which is stored under refrigeration, to room temperature and gently mix it.

2. Prepare three slides as follows: use a wax glass-marking pencil to draw an approximately 1-inch diameter circle on each of three glass slides. These circles are referred to as "wells." Label the three slides (wells):

   a. Patient serum

   b. Positive control

   c. Negative control (Fig. 18.2)

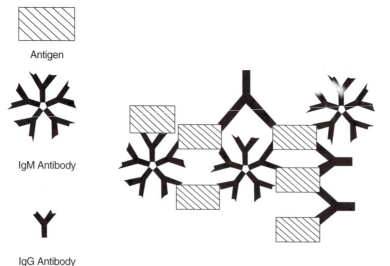

Antigen

IgM Antibody

IgG Antibody

FIG. 18.1. Agglutination reaction.

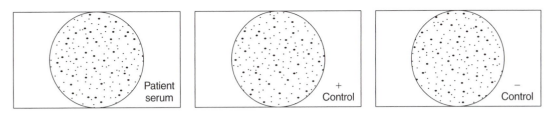

FIG. 18.2. Slides used for *Brucella* antigen test.

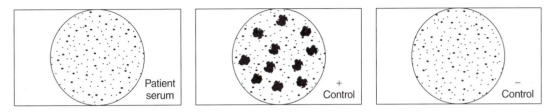

FIG. 18.3. Results of *Brucella* agglutination test.

3. Use a different 0.2 ml serological pipette to place 0.02 ml of each of the three sera into the appropriately labeled well. (See Fig. 18.2.) **Don't squeeze too much air out of the pipettes with the rubber bulb when drawing up fluid.** The solution should go only *slightly above* the 0.02 ml mark. The excess fluid should be expelled back into its original container by *gently* squeezing the rubber bulb. Alternatively, you may be instructed to place 1–2 drops of each of the three sera into the similarly labeled wells.

4. Add one drop of **mixed** *Brucella* antigen suspension to each of the three wells.

Note: Mix gently but thoroughly to ensure a smooth uniform suspension before adding the drops to the wells.

5. Use *separate* wooden applicator sticks (or toothpick) to mix and spread the contents over the entire surface of each well.

6. If a rotator is available, rotate the slides @ 180 rpm for 1 minute. If a rotator is not available, place the slides on a paper towel and shake the towel back and forth **carefully** for 1 minute to mix the solutions well. **Avoid spills; avoid turning slides over!** Should a spill occur, ask your instructor how to clean up before you touch anything.

7. Observe immediately—look for agglutination. (See Fig. 18.3.)

Results

The positive control must be agglutinated, and the negative control must be negative for agglutination. If the patient's serum is agglutinated, the test is presumptively positive. Further testing by a tube dilution method is required before a final diagnosis is made.

AN ENZYME-LINKED SEROLOGICAL TEST FOR RAPID IDENTIFICATION OF *STREPTOCOCCUS PYOGENES*

Materials List per Table/Work Station

Agar plate culture of *Streptococcus pyogenes*

⚠ **CAUTION:** NO AVIRULENT STRAINS OF THIS MICROBE ARE AVAILABLE FOR USE. IF NOT DONE AS A DEMONSTRATION, EXTREME CARE MUST BE USED.

Streptococcus pyogenes continues to be a major cause of acute pharyngitis, particularly in children. If not treated promptly, permanent damage to the heart, joints, and kidneys may occur in the form of rheumatic fever, rheumatoid arthritis, and glomerulonephritis. Traditional tests for this microbe frequently take 24 to 48 hours to complete (e.g., bacitracin sensitivity as covered in the previous exercise). Numerous serological tests are now available to rapidly detect *S. pyogenes* directly from a throat culture, thus reducing the time for diagnosis from a few days to a few minutes. The **Johnson & Johnson Sure Cell Strep A Test Kit®,** and many others, detect the presence of *S. pyogenes** directly from a patient's throat or from an isolated colony on a throat culture agar plate.

The basic principle of the test is as follows: suspected Group A strep antigen (the intact organism) is placed on a filter covered with antibody, which reacts with the antigen of Group A strep. If the strep antigen

---

**S. pyogenes* is also known as Group A strep based on a serological test developed by Dr. Rebecca Lancefield. For a more detailed explanation of Group A strep, see "Differentiation of Streptococci Using Latex Agglutination" later in this exercise.

is present, it becomes attached to the filter. If no antigen is present, no microbial attachment to the filter will occur. A second Group A strep specific antibody is then poured through the filter and adheres to the strep antigen that is present. Attached to this second antibody is the enzyme, horseradish peroxidase, which will cause a color change when the appropriate substrate is added. The only way the enzyme will cause a color change is if the antibody-enzyme conjugate attaches to the filter, and the only way the antibody is there is if Group A strep antigen is on the filter. Thus, a color change observed in this test indicates the presence of Group A antigen. (See Figs. 18.4–7.)

FIG. 18.4. Filter covered with anti-Group A strep antibody.

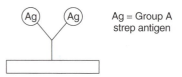

FIG. 18.5. Sample (from throat or growing colony), after pretreatment is added. If the Strep A antigen is in the sample, it attaches to the previously absorbed antibody.

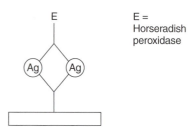

FIG. 18.6. A second antibody specific for the antigen is added. This second antibody has horseradish peroxidase enzyme attached to it. All enzyme-linked antibody not bound to antigen is washed away.

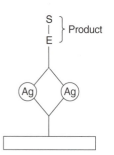

FIG. 18.7. When the substrate of the peroxidase enzyme is added, its reaction product causes a color change. This signifies that the test sample has the antigen. If there is no color change, the antigen is not present in the sample.

PROCEDURE

The procedure utilizes a negative control compartment as well as a positive control compartment.

Note: Regardless of the exact form of the Enzyme-Linked Immunosorbent Assay (ELISA) test utilized, the basic principle remains the same.

1. A sample of the suspected microbe is collected on a swab either directly from a patient or from culture media. The swab is rotated with a nitrite solution mixture in an "extraction block," which makes the *S. pyogenes* antigen available for the test.

2. A buffer solution is added to stabilize the reactants.

3. A sample of the buffered antigen is removed from the extraction block and placed into the center well of the test kit. If any *S. pyogenes* antigen is present, it will then adhere to the antibody attached to the filter already present in the well.

4. The second antibody is then added to all three wells. This second antibody has the enzyme horseradish peroxidase attached to it.

5. A wash solution is added to remove any unattached antibody-enzyme conjugate.

6. A dye solution (substrate) is added to all three wells. The dye will turn red or purple when it reacts with the enzyme, which is attached to the second antibody utilized in step 4 of the procedure.

DIFFERENTIATION OF STREPTOCOCCI USING LATEX AGGLUTINATION

The previous procedure is useful if *S. pyogenes* (Group A strep) is specifically suspected as the disease-causing or etiologic agent. The following procedure can be used to identify not only this specific microbe, but also several other categories of streptococci.

Nearly three-quarters of a century ago, Dr. Rebecca Lancefield discovered that streptococci contained unique antigenic carbohydrates on their cell surfaces. By reacting these carbohydrates with known specific antibodies, she was able to classify these microbes into six different groups: A, B, C, D, F, and G. For example, *Streptococcus pyogenes,* just covered, almost always has Antigen A on its surface. Therefore, the identification of Antigen A from a streptococcus specimen virtually identifies it as *S. pyogenes.*

Because of this reaction, *Streptococcus pyogenes* is also known as Group A strep, Group A beta-hemolytic strep, or Lancefield's Group A strep. The Lancefield grouping of streptococci is separate and individual from its type of hemolysis.

Slidex Strepto-KIT®

The Slidex Strepto-KIT (bioMérieux Vitek, Inc.) is a latex agglutination system for the rapid identification of the Lancefield grouping of beta-hemolytic streptococci (groups A, B, C, D, F, G). It is easy, rapid, and widely used clinically.

Streptococci possess group-specific antigens located on the surface of their cells. In the latex agglutination technique, the group-specific antigen is enzymatically extracted from the cell walls of isolated colonies or pure cultures of streptococci. Antigen in the enzyme extract is identified using latex particles conjugated to group-specific antisera (antibodies). Visible clumping, that is, *agglutination,* will form in the specific latex particle suspension that reacts with the specific extracted antigen. Conversely, the latex will remain in suspension if the antigen is not present in the enzyme extract.

Clinically, this serological test will be used on bacteria, which have been identified as Gram positive, beta-hemolytic, and catalase negative cocci, and are presumed to be streptococci.

Materials List per Table/Work Station

Pasteur pipettes

Inoculating loop

Timer

Microtubes

Sterile distilled water

Agar plate with isolated colonies/broth cultures of *S. pyogenes, S. agalactiae, E. faecalis*

Slidex Strepto-KIT:

six dropper bottles of latex streptococcal antiserum suspension (groups A, B, C, D, F, G)

extraction enzyme

positive control enzyme extracts of A, B, C, D, F, G

streptococci disposable cards

disposable stirring sticks

PROCEDURE

1. Students at each table should work as a group. Use the streptococcus species assigned to you.

Note: Since there are no avirulent strains of *S. pyogenes* available, it is suggested that the instructor demonstrate this reaction.

2. Obtain a small test tube of the extraction enzyme. Make sure the enzyme is at room temperature and not cold.

3. Using a loop, transfer four suspected streptococcus colonies into the small test tube of extraction enzyme. Mix well. Incubate the test tube for 10 to 15 minutes in the 37° C incubator. (If the strep is in broth, transfer 4 to 5 loopfuls.) (See Fig. 18.8.)

4. When incubation is completed, do not immediately remove the extraction enzyme from the incubator. First do steps 5–7.

5. Obtain a Strepto-KIT card. Do not touch the reaction areas of the card.

6. Obtain the six dropper bottles of latex streptococcal antiserum suspension (groups A, B, C, D, F, G). Shake each bottle well. Make sure they are warmed to room temperature.

7. Dispense 1 drop from the dropper bottle of sensitized latex reagent A into square A of the card. Dispense reagent B into square B; reagent C into square C; and so on, until each reagent has been added to its respective square on the card. (See Fig. 18.9.)

8. Now remove the extraction enzyme from incubation. Using a small Pasteur transfer pipette, dispense 1–2 drops of the incubated enzyme into EACH of the six squares on the Strepto-Card. Take care not to touch the pipette to any one square. (See Fig. 18.10.)

9. Using a stick or toothpick, mix the contents of square A. Spread the contents across the entire square. Using a NEW stick for each square, mix and spread the contents of the remaining squares.

10. Wait 2 minutes. Agglutination (clumping) will take place in the square containing the antibody against the streptococcus antigen. The other squares will not agglutinate.

11. Determine in which Lancefield group your assigned species belongs. Dispose of the cards in the biohazard bin.

Results

S. pyogenes—Group A

S. agalactiae—Group B

E. faecalis—Group D

(The other groups are associated with animals, not humans.)

PROCEDURE CONTROLS

Prior to the use of the Slidex Strepto-KIT, the following control measures should be conducted by one member of the class:

1. *Positive Control:* Using 1 drop of the positive control provided in the kit in place of isolate extract, follow steps 5 through 9. Each latex suspension should show strong agglutination results.

2. *Negative Control:* Without adding any organism to the extraction enzyme, follow all steps of the procedure. No agglutination should be seen in any of the latex suspensions.

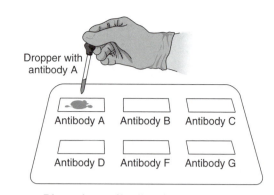

FIG. 18.9. Place the antibodies for each group of streptococci onto the appropriate squares on the Strepto-Card antibody A to square A, antibody B to square B and so on.

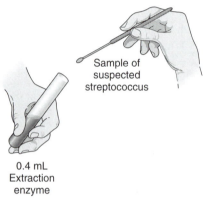

FIG. 18.8. Mix a loopful of the suspected streptococcus in 0.4 ml extraction enzyme to release the antigen.

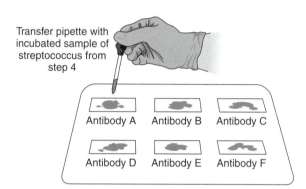

FIG. 18.10. Place 1–2 drops of the streptococcus-extraction enzyme mixture into each of the six squares on the Strepto-Card.

NAME _____ DATE _____ SECTION _____

QUESTIONS

1. Differentiate between blood plasma and serum.

2. What population of people is at greatest risk of suffering from Brucellosis?

3. What other agglutination reaction (besides the one that demonstrates Brucellosis) can you describe?

4. Discuss the necessity of using a positive and negative control in assessing agglutination and enzyme linked procedures.

5. What are "febrile" antigens?

6. How do IgM and IgG cause agglutination?

7. Name the etiologic agent that causes rheumatic fever.

8. Name the enzyme linked to the second Group A strep-specific antibody.

9. Explain the function of the linked enzyme in ELISA.

10. Which serious disease, caused by a retrovirus, is usually screened for by an ELISA technique?

11. Describe the mechanism of the reactions studied with the immunologic tests described in this exercise.

12. Distinguish between qualitative and quantitative serologic results.

13. Identify the Lancefield group for *S. pyogenes* and *E. faecalis.*

14. What is the basis for Lancefield's assignment of different species of streptococcus into groups?

15. Name one species of streptococcus associated with each of the different Lancefield groups known to infect humans.

16. Does any species of streptococcus not have a Lancefield group?

MATCHING

a. serology

b. serum

c. *Brucella*

d. agglutination

e. plasma

f. ELISA

g. Group A strep

h. Lancefield's carbohydrate classification

_____ liquid portion of blood containing clotting factors

_____ clumping action due to reaction between antibody, antigen (and often latex particles)

_____ *Streptococcus pyogenes*

_____ serological means of identifying streptococci

_____ uses a peroxidase enzyme to identify a suspected antigen

_____ process by which the presence of disease can be detected by mixing antibody and antigen

_____ disease entity identified by the presence of a febrile antigen

MULTIPLE CHOICE

1. Which of the following is used to detect the presence of an antigen by way of an enzyme labeled antibody?

   a. Landsteiner's procedure   b. agglutination   c. ELISA   d. Lancefield's antibody

2. Brucellosis is a disease associated with which of the following professions?

   a. respiratory care specialists   b. veterinarians   c. physical therapists   d. radiation technologists

3. Antibody is attached to latex beads. When these antibodies are reacted with an antigen specific for that antigen:

   a. the reaction heats up   b. clumping occurs   c. a dark color appears   d. the antigen dissolves

4. The study of the interactions of antibody and antigen, especially when used to identify disease entities, is:

   a. serology   b. bacteriology   c. rheumatology   d. hematology

5. *Streptococcus pyogenes* has which type of carbohydrate antigen on its capsule?

   a. b   b. d   c. a   d. c

Agglutination Sticking together of microscopic antigen and antibody particles.

Antibody An immunoglobin produced in the body which binds specifically to an antigen that causes the antibody to be produced.

Avirulent A microbe or strain of a microbe that has few, if any, dangerous properties, for example, resistance to antimicrobials, protective capsule, and ability to produce toxins.

ELISA—Enzyme Linked Immunosorbent Assay A known antigen that was attached to a plastic surface reacts with (i.e., binds itself to) the antibody tested for. A secondary antibody, to which an enzyme is attached (linked—called a conjugate), now is added and complexes to the original antigen-antibody complex. After each of the previous substances is added, all unattached (uncomplexed) substances are washed away. A substrate is then added, which the linked enzyme converts into a colored product. The amount of visible color is proportional to the concentration of the original antibody from the patient. Therefore, this is a quantitative test of antibody concentration or titer.

Endemic A disease entity (etiologic agent) that is always present within a population or geographical location.

Epitope A component of an antigenic molecule that reacts with an antibody.

Extraction enzymes Enzymes used in serological testing to cleave surface antigens from the microbial cell.

Febrile antigen Any species of pathogenic microbe that causes fever in infected individuals.

Immunology Study of how the immune system reacts to stimulation by specific infectious organisms.

Lancefield groupings Species of streptococcus divided into individual groups based on surface antigens.

Multivalent (also polyvalent) Antibodies capable of reacting with more than one strain or type of specific antigen or organism.

Plasma The liquid portion of blood, including the clotting factors.

Secondary antibody An antibody that will bind to another antibody to complete an assay procedure such as ELISA.

Serology Science which employs serum to detect antigens and antigens to detect antibodies in serum.

Serum The liquid portion of plasma without the clotting factors.

EXERCISE 19

Identification of Enteric Pathogens: Traditional Methods

Objectives

After completing this lab, you should be able to:

1. Distinguish between bacteria that belong to the family *Enterobacteriaceae* versus *Non-Enterobactiaceae.*

2. State the significance of glucose fermentation and the oxidase test in identifying Gram negative rods.

3. Explain the procedures for the IMViC tests.

4. Accumulate information through biochemical tests that will allow you to eventually identify your bacterial unknown(s).

Once a microbiology laboratory receives bacterial specimens, it usually has to perform two procedures known as *culture and sensitivity* tests. Both rely on well-isolated colonies. The *culture* refers to the process of growing the microbe on media appropriate for identifying the etiologic or causative agent of the disease. The sensitivity procedure (e.g., Kirby-Bauer technique) determines which drugs the microbe is sensitive to. Bacteria causing GI tract infections were once considered the most difficult to identify because most of the etiologic agents* are Gram negative rods and are virtually indistinguishable from each other under the microscope.

Various tests were developed to help the microbiologist determine which specific microbe caused which specific disease. Identification of the microbe is important to aid in predicting and preventing further outbreaks or epidemics, whether they be in the hospital, city, county, or nation.

This exercise concerns itself with the more traditional biochemical procedures used for identification.

*The microbe that causes a particular disease.

When these tests were originally developed, typhoid fever (*Salmonella typhi*), bacillary dysentery (*Shigella dysenteriae*), and cholera (*Vibrio cholera*) were common serious etiologic agents of the GI tract in the United States. (These microbes remain a serious threat in other parts of the world.) The tests originally devised for these microbes are still used today to identify other Gram negative enteric pathogens.

The first step (after isolating colonies) in this procedure is to determine which major group of Gram negative rods you are dealing with. One major group is the *Enterobacteriaceae* (*entero* = intestinal, *-aceae* = family). This group has three common characteristics.

1. They are all Gram negative.

2. They all ferment glucose.

3. They test negative for the enzyme *oxidase*. (See Exercise 17 on specific laboratory tests for a review of this last procedure.)

The *Enterobacteriaceae* are also known as "Enteros" and "Fermenters."

TABLE 19.1	**DIFFERENTIATION OF GRAM NEGATIVE RODS**	
	Enterobacteriaceae (*Fermenters, enteros*)	*Non-Enertobacteriaceae* (*Non-Fermenters, Non-Enteros*)
	Ferments glucose	Does not ferment glucose
	Oxidase negative	Often oxidase positive

The other major group of Gram negative rods are comprised of several different families and genera of bacteria and are collectively known as the *Non-Enterobacteriaceae* (also called Non-Enteros and Non-Fermenters). This group also has three common characteristics.

1. They are all Gram negative.

2. They do not ferment glucose, but they may oxidize it.

3. Many, but not all, are oxidase positive.

There should, therefore, be no surprise that glucose fermentation and the oxidase test are part of this exercise. (See Table 19.1.)

Materials/Student (if both unknowns are assigned, double the amount of materials)

Isolated colonies of unknown from previous exercise or unknown culture provided

One T-Soy Agar plate

One Triple Sugar Iron agar

One Sulfide-Indole-Motility medium

Two O.F. basal medium tubes with glucose

One to three tubes of decarboxylase broth with lysine/ornithine/arginine

One tube of sterile mineral oil

One tryptic-nitrate medium tube

One IMViC test: One T-Soy broth for indole production (may be omitted as this reaction is done in the SIM tube); two MR-VP broths—5 mls of broth per tube; one Simmons citrate slant

One tube of nutrient gelatin

Materials/Per Table/Work Station

Container of oxidase test reagent

Several toothpicks, swabs, or plastic inoculators

Filter paper or white paper towels

PROCEDURE

Observe your isolated colonies from Exercise 16. Choose several well-isolated colonies with the same cultural characteristics. Use these colonies for all inoculations. (An unknown culture may be given to you for these tests.)

If you have no isolated colonies, restreak a portion of your unknown on the T-Soy plate. You will need isolated colonies to continue the identification procedure in a future lab. Inoculate the media listed below with your mixed culture for practice and review, but you cannot use the results solely for identification purposes.

1. *Inoculation of T-Soy Agar.* Touch one well-isolated colony and streak it out on the T-Soy agar. This plate will be your source of inoculum for the next lab and will confirm that you isolated the sample correctly. (*Optional:* You might also be instructed to inoculate an Eosin-Methylene-Blue or MacConkey plate.)

2. *Inoculation of Triple Sugar Iron Agar.* Review Exercise 13 for the inoculating procedure and function of this medium; then inoculate one tube with your unknown.

3. *Inoculation of Sulfide-Indole-Motility Medium.* Review Exercise 13 for the inoculating procedure and function of this medium; then inoculate one tube with your unknown.

4. *Inoculation of O.F. Basal Medium with Glucose.* Review Exercise 11 for the inoculating procedure and function of this medium; then inoculate two tubes with your unknown. (Remember the mineral oil.)

5. *Inoculation of Decarboxylase Broth with Amino Acid(s).* Review Exercise 11 for the inoculating procedure and function of this medium; then inoculate one to three tubes with your unknown.

6. *Inoculation of Tryptic-Nitrate Medium.* Review Exercise 11 for the inoculating procedure and function of this medium; then inoculate one tube with your unknown.

7. *Inoculation of the IMViC Set of Reactions.* The differential media just listed have been proven useful in identifying Gram negative rods. To a certain extent, they help group the hundreds of Gram negative rods of the *Enterobacteriaceae* and *Non-Enterobacteriaceae* in smaller, easier-to-manage categories. Once in these smaller categories, identification of specific microbes is somewhat easier. The development of the IMViC tests, however, enabled the early microbiologists to readily identify specific microbes directly. For example, certain strains of *E. coli* and *Enterobacter aerogenes* share many cultural characteristics and grow alike on both MacConkey and Eosin-Methylene Blue agar (they both ferment lactose). The IMViC test readily distinguishes be-

tween them, as it does between most strains of *E. coli* and *Klebsiella pneumoniae*.

The IMViC test actually comprises four different biochemical tests:

I = Indole. Indole production is from tryptophan. See Exercise 11 for a review of the inoculating procedure and function of this test which uses T-soy broth.

M = Methyl Red. Methyl red is a pH indicator that turns red in extremely acidic solutions (pH4–5) and yellow in less acidic environments. The differential medium for this procedure is *MR-VP broth*. Methyl red is rather toxic to bacteria, so it must be added to the MR-VP broth after growth has occurred. Otherwise, its presence may inhibit bacterial growth, unlike the other pH indicators used in this course.

V = Voges-Proskauer. The Voges-Proskauer test, named for its developers, determines the presence of the chemical byproduct *acetoin* or *acetyl methyl carbinol* from glucose metabolism. The medium used to test for this phenomenon is also MR-VP broth. Potassium hydroxide (with or without creatine) and α-naphthol are usually the reagents used to detect the presence of acetoin. A red color, which may take up to 30 minutes to develop, is indicative of a positive reaction.

i = no test. The lowercase "i" is added to aid in pronunciation of the test.

C = Citrate. See Exercise 11 for a review of the inoculating procedure and function of Simmon's citrate medium.

Note: The MR-VP medium will require at least 48 hours of incubation before reagents are added and results read.

8. *Inoculation of Nutrient Gelatin.* Hydrolysis or liquification of gelatin can be used to distinguish *Serratia* and *Pseudomonas* species from other bacterial groups. By stabbing an inoculum into a *nutrient gelatin deep,* the presence of the enzyme for gelatin catabolism can be readily determined. The tube is initially solid due to the presence of gelatin and not agar. After incubation, the tube is cooled to room temperature or refrigerated, and it can be read by carefully tilting the tube. If the medium in the tube is solid, the bacteria did not hydrolyze or liquify the gelatin. If it is liquid, the reaction is positive. If the room is extremely warm, or if the culture is fresh from the incubator, the tubes may have to be cooled in an ice bath before they are read.

Note: Nutrient gelatin often requires a 48-hour incubation period for accurate results.

9. *Oxidase Test.* The oxidase test is used to distinguish *Enterobacteriaceae* (negative) from *Non-Enterobacteriaceae* (most are positive). The reagent, (tetramethyl-para-phenylenediamine dihydrochloride), turns color in the presence of cytochrome oxidase enzymes, which are the same enzymes found in the electron transport system of human cells. A purple color indicates the presence of these oxidase enzymes.

PROCEDURE

Inoculate all the assigned media and place in the incubation tray. Stab the nutrient gelatin tube as directed by your laboratory instructor. Separate the MR-VP and nutrient gelatin tubes for a 48-hour incubation.

Use a paper towel or piece of filter paper to perform the oxidase test on your unknown. Place one drop of the oxidase reagent on the paper towel or filter paper. Aseptically remove a well-isolated colony from the T-Soy plate, using the end of a toothpick, plastic inoculator, or handle of a cotton swab. Do not use a metal loop because certain types of metal will often give a false positive reaction. A purple color, which must develop within *30 seconds,* indicates the presence of oxidase enzymes. Any color change after this time must be considered negative. If you do not have isolated colonies, perform the test anyway for practice. Repeat the test if necessary once you achieve colony isolation.

Results

Reagents required to test for reactions

Solution A and Solution B—tryptic nitrate

Zinc—tryptic nitrate

Kovac's reagent—indole production

Methyl red—methyl red acidity test

Potassium hydroxide—with or without creatine—Voges-Proskauer test

α-Naphthol—Voges-Proskauer Test

T-Soy Plate. Inspect the plate and confirm that all colonies display the same cultural characteristics, which indicates a pure culture. If more than one colonial type is seen, it may mean that the samples used for the inoculations were contaminated. If you were not able to isolate colonies initially, determine whether you were more successful at the second attempt.

Triple Sugar Iron Agar, Sulfide-Indole-Motility, O.F. Basal, Decarboxylase Broth(s), Tryptic Nitrate. Refer to Exercises 11 and 13 for a review of which reagents to use as well as an interpretation of the results.

IMViC TEST

Indole: Add several drops of Kovac's reagent to the T-Soy broth tube and determine whether indole has been produced. Remember to use the fume hood. Refer to Exercises 11 and 13 for a review of how to interpret the results.

Methyl Red: Add 5 drops of methyl red to one of the MR-VP tubes and mix gently. A red color indicates the presence of a high concentration of acid (positive), and a yellow color indicates a negative reaction.

Voges-Proskauer: Add approximately 1 ml of 5% α-naphthol and approximately half that amount of 40% potassium hydroxide to the other MR-VP tube. A red color indicates the presence of acetoin. The color change may take up to 30 minutes to develop. (An alternative procedure is to aseptically remove 1 ml of the broth, then add 0.6 ml of the 5% α-naphthol, followed by 0.2 ml of 40% potassium hydroxide. This higher ratio between the broth and the reagents usually gives a result within 5 minutes but necessitates pipetting a broth culture of bacteria.)

Citrate: Observe the slant of the Simmons citrate tube for a color change. Refer to Exercise 11 for a review of how to interpret the results.

Gelatin Hydrolysis: Carefully tilt the tube of nutrient gelatin and determine whether the bacteria hydrolyzed (liquified) the gelatin. If the cultures came right from the incubator or if the room is very warm, the tubes may have to be refrigerated for several minutes first.

Record the results of the oxidase test in Table 19.2.

Fill in Table 19.2 with the reactions of your unknown.

Inventory

At the end of this laboratory, each person will have the following ready for incubation: (If both unknowns are to be identified, double the number of plates and tubes.)

One T-Soy agar plate

One Triple Sugar Iron agar slant

One Sulfide-Indole-Motility deep (can be used to substitute for part of the IMViC test)

Two O.F. basal medium deeps with glucose (remember the mineral oil)

One to three tubes of decarboxylase broth with lysine/ornithine/arginine

One tryptic-nitrate tube for the nitrate reduction test

One set of IMViC reaction tubes: One T-Soy broth for indole production test; two MR-VP broths; one Simmons citrate slant

One nutrient gelatin deep

TABLE 19.2

Biochemical Test		Reaction(s)			
Triple Sugar Iron agar		**Glucose**	**Lactose**	**Gas from Glucose**	**H$_2$S**
Sulfide-Indole-Motility		**H$_2$S**	**Indole**	**Motility**	
O.F. Basal with Glucose		**With oil**		**Without oil**	
Decarboxylase Broth	**Lysine**				
	Ornithine				
	Arginine				
Tryptic-Nitrate Broth (Nitrate Reduction)					
IMViC	**Indole**				
	Methyl-Red				
	Voges-Proskauer				
	Citrate				
Gelatin Hydrolysis					
Oxidase Test					

NAME _____ DATE _____ SECTION _____

QUESTIONS

1. Which two procedures are usually performed on specimens sent to the microbiology lab?

2. Name two biochemical tests used to differentiate between the *Enterobacteriaceae* and *Non-Enterobacteriaceae*.

3. Which inoculation procedure(s) performed in this exercise will aid in classifying your unknown as an *Enterobacteriaceae* or a *Non-Enterobacteriaceae?*

4. Outline the steps of the oxidase test as performed in class. Why can't a wire loop be used to perform this test?

5. Fill out the following chart for the IMViC test:

Test	Type of* Medium	Reaction Tested for	+/− Reaction (Color)	Reagent(s) needed
Indole				
Methyl red				
Voges-Proskauer				
Citrate utilization				

*Selective, differential, etc.

MATCHING

a. oxidase negative, glucose
   fermentation positive

b. Triple Sugar Iron agar

c. Sulfide-Indole-Motility
   medium

d. O.F. basal with glucose

e. Decarboxylase broth

f. Kovacs reagent

g. Methyl red

h. Voges-Proskauer test

_____ tests for utilization of iron salt, catabolism of tryptophan, and presence of
flagella

_____ test for removal of acid from an amino acid

_____ reagent used to test for the presence of indole

_____ family *Enterobacteriaceae*

_____ tests for glucose, lactose, and sucrose fermentation, gas production, utilization of iron salt.

_____ test for fermentation or oxidation of glucose

_____ test for the presence of acetoin

MULTIPLE CHOICE

1. A MacConkey agar plate has been inoculated with an unknown bacterium. There is growth with a pink color. The
   microbe is a(n):

   a. *Enterobacteriaceae*   b. Gram positive   c. Gram negative   d. decarboxylase negative

2. A microbe was inoculated into a tube of decarboxylase broth. A negative reaction is indicated by which color?

   a. yellow   b. red   c. blue   d. green

3. A positive reaction in Simmons citrate is indicated by which color?

   a. yellow   b. red   c. blue   d. green

4. The methyl red test is used to indicate the presence of:

   a. acetoin   b. indole   c. gelatin hydrolysis   d. acid

5. O.F. basal medium tubes with glucose were inoculated with a test microbe. After 24 hours of incubation, both
   tubes were green. This indicates:

   a. fermentation   b. oxidation   c. negative reaction   d. both oxidation and fermentation

6. The substrate in the decarboxylase test is:

   a. iron salts   b. amino acids   c. gelatin   d. acetoin

7. A tube of tryptic-nitrate broth was inoculated with a pure culture. After 24 hours, Solution A and Solution B were
   added. There was no color change. Which of the following statements is true?

   a. this is a test for nitrate reduction   b. nitrite is not present
   c. the addition of zinc is the next step   d. all of these

8. Which statement is false about Triple Sugar Iron agar?

   a. red slant = lactose negative   b. yellow butt = glucose positive
   c. black color = acetoin production   d. cracks/bubbles = gas production

WORKING DEFINITIONS AND TERMS

Enterobacteriaceae Family of facultative anaerobic Gram negative rods traditionally associated with intestinal flora, such as *E. coli,* and intestinal diseases, such as typhoid fever and dysentery.

IMViC Set of four reactions developed to readily identify certain Gram negative rods by testing for indole production, acidity, presence of acetoin, and citrate utilization.

Non-Enterobacteriaceae Virtually any facultative anaerobic Gram negative rod not in the family *Enterobacteriaceae.* Most are oxidase positive.

Oxidase test (Test for cytochrome oxidase) Any of a group of enzymes that function as part of the electron transport system found in human cells and certain bacterial cells. The *Enterobacteriaceae* do not possess this enzyme, whereas most of the *Non-Enterobacteriaceae* do.

Identification of Enteric Pathogens: Rapid Identification Methods

Objectives

After completing this lab, you should be able to:

1. State which type of Gram negative bacteria the Enterotube™ II is used to identify.

2. State which type of Gram negative bacteria the Oxi/Ferm™ Tube II is used to identify.

3. State which type of Gram negative bacteria the API® 20 E System is used to identify.

4. Correctly perform the procedures for inoculating the Enterotube™ II, the Oxi/Ferm™ Tube II, and the API® 20 E System.

5. Interpret the results of the inoculations made to these rapid identification systems, using appropriate references.

In the previous exercise, you were introduced to the more traditional testing methods for identifying many Gram negative rods. Those tests, coupled with the rapid, mini-identification systems that will be utilized in this lab exercise, will provide you with enough information to identify your assigned unknown. Accuracy will depend on whether a pure culture was used, the media inoculated and prepared for incubation properly, the correct reagents added, and results read properly. The media used in this exercise are meant for commercial use in hospitals and clinics and are designed to be read or interpreted immediately after incubation. In most cases, the test media will be refrigerated after 24 hours of incubation, which may affect the accuracy of some of the individual tests.

ENTROTUBE™ II

The Enterotube™ II is a multimedia tube containing 12 separate compartments of different media, which tests for a total of 15 different biochemical reactions. It is used primarily on Gram negative rods suspected of belonging to the family *Enterobacteriaceae* (oxidase negative). (Refer to Table 19.1 for a review of Enteros and Non-Enteros and also see Fig. 20.1.)

The following is a synopsis of the biochemical tests contained in the Enterotube™ II. A more detailed explanation of the reactions and procedures is available from the manufacturer's pamphlet and instruction guide.

Compartment 1—Glucose Fermentation and Gas Production

This compartment contains glucose and the pH indicator, phenol red. A layer of wax makes the environment anaerobic (fermentation) and will indicate whether gas was produced by its separation from the agar. A color change from red to yellow indicates that the glucose was fermented.

Compartments 2 and 3: Lysine and Ornithine Decarboxylation

These two chambers are also covered with wax, for the decarboxylase reaction must be observed under anaerobic

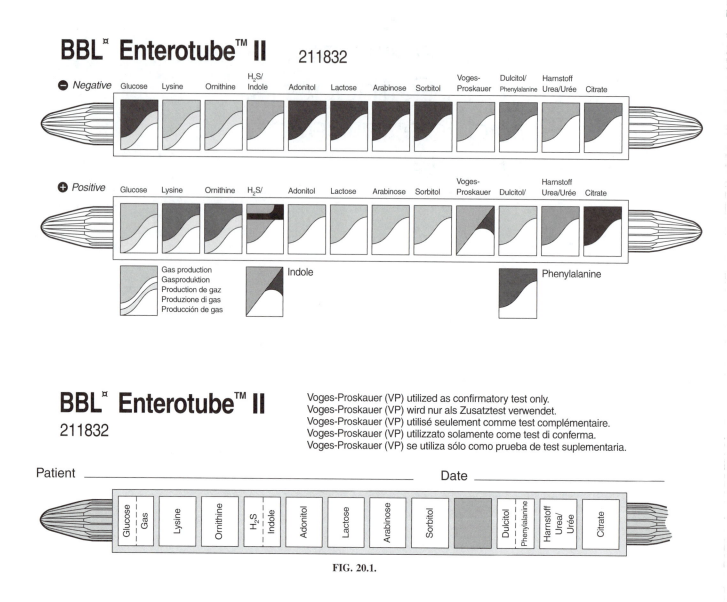

FIG. 20.1.

conditions. The reaction seen is the same as the decarboxylase broth previously used, except that the pH has been adjusted. As a result, the color of these compartments is yellow rather than purple as in the original broth tubes used. A color change from yellow to purple indicates a positive reaction.

Compartment 4: Sulfide and Indole Production

This compartment basically contains the same formula as the Sulfide-Indole-Motility medium. However, motility cannot be tested for in this form because the increased concentration of agar used to maintain the integrity of the medium prevents detection of motility. A blackening of the medium indicates sulfide production, and a red color resulting from the addition of Kovac's reagent means that indole is present.

Compartments 5–8: Adonitol, Lactose, Arabinose, and Sorbitol Fermentation

Each compartment contains the respective sugar with phenol red pH indicator. Each is read the same as the phenol red fermentation tubes as in earlier laboratories, except the gas production cannot be determined. A color change from red to yellow indicates a positive reaction.

Compartment 9: Voges-Proskauer

This chamber tests for the presence of acetoin as a byproduct of glucose metabolism. A red color in the presence of α-naphthol and potassium hydroxide indicates a positive reaction.

Compartment 10: Phenylalanine Deaminase and Dulcitol

The *phenylalanine deaminase* test detects the presence of *pyruvate*, which is a byproduct of the deamination of the amino acid phenylalanine (as the name of the test indicates). Pyruvate turns a dark brown, gray, or black color in the presence of ferric chloride, $FeCl_3$, which is mixed with the medium. Since the original medium is colorless, this color change will be obvious.

Dulcitol is a carbohydrate. The pH indicator in this chamber turns yellow when this sugar is utilized.

After incubation, one of three different reactions will be seen in this compartment:

Phenylalanine positive: Agar will be dark brown, black, or gray

Dulcitol positive: Agar will be yellow

Both negative: Agar will be a light green color

Note: None of the bacteria tested will be positive for both Phenylalanine Deaminase and Dulcitol

Compartment 11: Urea

This chamber tests for the presence of the enzyme urease. A color change from light pink to purple or dark pink indicates a positive reaction.

Compartment 12: Citrate

This chamber tests for citrate utilization as the only carbon source for energy. The color change from green to blue is indicative of a positive reaction as seen in previous exercises.

OXI/FERM™ TUBE II

The Oxi/Ferm™ Tube II is used when the Gram negative rod is suspected of being a *Non-Enterobacteriaceae* (which are usually oxidase positive). As with the Enterotube™ II, there are 12 compartments, which are used to perform 14 biochemical tests. The tube's name is derived from the fermentation and oxidation of glucose (O.F. basal medium with glucose). See Fig. 20.2

The following is a synopsis of the biochemical tests contained in the Oxi/Ferm™ Tube II. A more detailed explanation of the reactions and procedures is available from the manufacturer's pamphlets and instruction guide.

Compartment 1: Anaerobic Glucose

This compartment tests for glucose fermentation and uses a layer of wax to provide the anaerobic environment. A color change from green to yellow indicates a positive reaction.

Compartment 2: Arginine Dihydrolase

This compartment tests for a decarboxylase (also called dihydrolase) reaction using the same pH indicator as previously utilized. A color change from yellow to purple indicates a positive reaction.

Compartment 3: Lysine

This compartment tests for a decarboxylase reaction using the same pH indicator as previously utilized. A color change from yellow to purple indicates a positive reaction.

Compartment 4: Lactose Fermentation and Nitrogen Gas Production (Lactose/N_2)

This compartment tests for lactose fermentation as well as reduction of nitrate to nitrogen gas. Lactose fermentation is detected by a yellow color in the medium. Red or gray is negative. Nitrogen gas production is determined by observing any separation of the wax from the agar or any separation of the agar from the compartment wall.

Compartment 5: Sucrose Oxidation and Indole Production

Oxidation of sucrose is detected by a color change from green to yellow (as in the O.F. basal medium). Indole is detected by the addition of Kovac's reagent to this compartment.

Compartments 6–9: Xylose, Aerobic Glucose, Maltose, and Mannitol Oxidation

These compartments contain O.F. basal medium. A yellow color indicates that the carbohydrate has been oxidized. Green or blue indicates a negative reaction.

Compartment 10: Phenylanaline Deaminase

This test is the same as the phenylalanine deaminase test in the Enterotube™ II test procedure. When this test is read in the Oxi/Ferm™ Tube II, any shade of brown is considered positive.

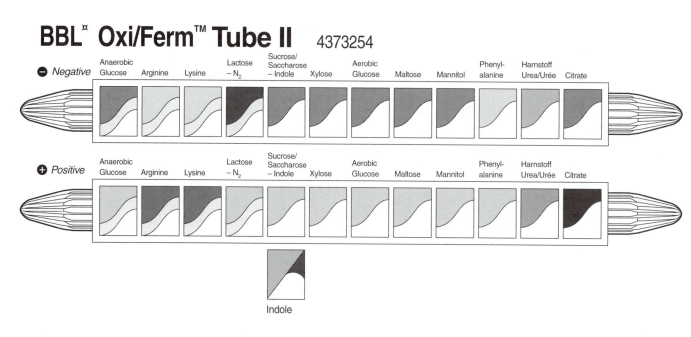

BBL¤ Oxi/Ferm™ Tube II
4373254

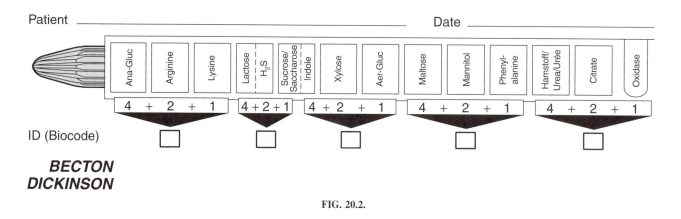

FIG. 20.2.

Compartments 11 and 12: Urea and Citrate

These tests are the same as those for the urea and citrate in the Enterotube™ II test procedure.

API® 20E SYSTEM

The API® 20 E System (API = Analytical Profile Index) performs at least 20 different biochemical tests of a suspected enteric bacterium. The tests are performed on a paper strip with 20 "microtubes," which is then placed in its own individual incubation chamber. (See Fig. 20.3.) Each microtube contains a specific substrate and

nothing else, except possibly a pH color indicator. A single substrate is incapable of supporting bacterial growth; thus, reactions are based on the bacterial enzymes placed in each microtube with the inoculum. Since growth does not occur, certain aspects of aseptic technique do not have to be adhered to when the test strip is prepared. (Note that the strip of microtubes is not kept covered.) Although intended primarily for the identification of *Enterobacteriaceae,* this system of tests also has the advantage of being able to identify many Non-Enteros.

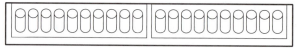

FIG. 20.3. Test strip for API 20 E®

As is true of the Enterotube™ II and Oxi/Ferm™ Tube II described earlier, many of the tests are principally the same as others performed in previous exercises. The following is a synopsis of the biochemical tests contained in the API® 20 E System. A more detailed explanation of the reactions and procedures is available from the manufacturer's pamphlet and instruction guide.

Microtube 1: Ortho-Nitrophenyl-Beta-D-ortho-nitrophenyl-β-D-galactopyranoside (ONPG)

The test determines the presence of the enzyme beta-galactosidase. A positive reaction is signified by a light yellow color.

Microtubes 2–4: Arginine Dihydrolase (DECARBOXYLASE), Lysine Decarboxylase, and Ornithine Decarboxylase (ADH, LDC, ODC)

These three microtubes test decarboxylase reactions. A positive reaction is signified by a color change from yellow to red.

Microtube 5: Citrate (CIT)

This microtube tests for citrate as the only carbon source for energy. A positive reaction is signified by a color change from green to blue.

Microtube 6: Sulfide (H₂S)

This microtube tests for sulfide production. A black precipitant is indicative of a positive reaction.

Microtube 7: Urea (URE)

This microtube tests for the enzyme urease. A dark pink or purple color is indicative of a positive reaction.

Microtube 8: Tryptophan Deaminase (TDA)

This test is virtually the same as the phenylalanine deaminase (PA) reaction of the Enterotube™ II. It measures the formation of pyruvic acid, which turns dark brown from colorless or light brown in the presence of ferric chloride (FeCl₃). The ferric chloride is premixed in the compartment testing PA in the

Enterotube™ II, but must be added after incubation in the API® 20 E System.

Microtube 9: Indole (IND)

Indole production from tryptophan is detected once Kovac's reagent is added. The formation of a red ring is indicative of a positive reaction, as it is in all the other indole tests performed previously.

Microtube 10: Voges-Proskauer (VP)

This microtube tests for the catabolism of glucose into acetoin. A red color is indicative of a positive reaction once α-naphthol and potassium hydroxide are added.

Microtube 11: Gelatin Hydrolysis (GEL)

This microtube determines whether the enzyme for gelatin hydrolysis (liquification) is present. The chamber contains powdered charcoal wrapped in a gelatin "envelope." If gelatin is hydrolyzed, the charcoal will diffuse evenly throughout the microtube.

Microtubes 12–20: Utilization of the Carbohydrates Glucose, Mannitol, Inositol, Sorbitol, Rhamnose, Saccharose (Sucrose), Melibiose, Amygdalin, and Arabinose (GLU, MAN, INO, SOR, RHA, SAC, MEL, AMY, and ARA)

A color change from blue to yellow indicates a positive reaction in each microtube.

In addition to these 20 tests, nitrate reduction and the catalase reaction can be performed by adding the appropriate reagents to the carbohydrate microtubes.

INOCULATION OF RAPID IDENTIFICATION MEDIA

Materials/Person

One Enterotube™ II

One Oxi/Ferm™ Tube II

One API® 20 E System consisting of a plastic tray, plastic cover, and test strip with 20 microtubes

One disposable Pasteur pipette

One 5 ml tube of sterile saline

Materials List per Table/Work Station

Mineral oil

INOCULATION OF THE ENTEROTUBE™ II

1. Remove the caps from both ends of the tube. The looped wire under the blue cap will act as your handle. The straight end under the white cap will be used to touch the bacterial sample. (Do not flame the straight end.)

2. Hold the tube as you would a pencil or an inoculating loop.

3. Touch an isolated colony with the straight end, twist the wire by turning the looped end, and then pull the wire completely through the tube. See Figs. 20.4 and 20.5.

4. Re-insert the wire through all the compartments so that the tip of the wire is just inside the citrate test chamber. Once completed, bend and break the wire. There is a notch on the wire, which will make it easy to break.

⚠ CAUTION: IF THE WIRE IS PUSHED TOO FAR INTO THE CITRATE COMPARTMENT, THE NOTCH WILL BE INSIDE THE TUBE AND THE WIRE WILL NOT BREAK!

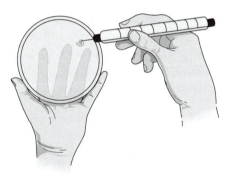

FIG. 20.4. Acquiring sample for Enterotube® II and Oxi/Ferm® Tube II.

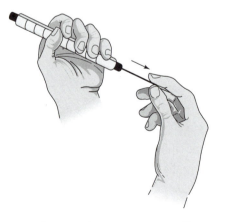

FIG. 20.5. Inoculation of Enterotube® II and Oxi/Ferm® II.

5. Screw the caps back on the ends of the tube.

6. Use the remnant of the inoculating wire handle to punch holes into the last eight chambers of the tube, thus exposing them to air. If the tube is rotated so that the labeled side is facing away, tape-covered notches can readily be seen in the upper region of these eight chambers. Once holes are punched into each of these chambers, the inoculating wire can then be discarded in the regular trash.

7. Label the tube with your name, date, and unknown number.

INOCULATION OF THE OXI/FERM™ TUBE II

Follow the same procedure as for the Enterotube™ II. The looped wire that will act as your handle is under the red cap.

INOCULATION OF THE API® 20 E SYSTEM

1. Place approximately 5 ml of water in the plastic tray, and place the test strip with the 20 microtubes in the tray.

2. Inoculate a sample of your unknown in a tube of sterile saline using a loop, Pasteur pipette, or a sterile swab. Mix until the saline is *slightly* cloudy.

3. Draw up several ml of the saline solution into a disposable Pasteur pipette and fill the bottom portion of all 20 of the microtubes. This is best accomplished by holding the tray at a 30–45° with the table, touching the Pasteur pipette to the side of the opening, and squeezing the correct amount of the inoculum into each chamber. (See Fig. 20.6.)

4. Once the bottom of each microtube is filled, place the tray flat on the table again. Completely fill the

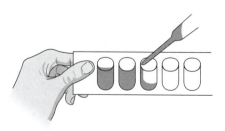

FIG. 20.6. Inoculation of API 20 E®.

open, upper parts of the microtubes (called the *cupule*) that have labels of the substrates bracketed. That is, completely fill the **[CIT],[VP],** and **[GEL]** microtubes.

5. Add several drops of mineral oil to the cupule portion of the microtubes that have the labels of the substrates *underlined*. That is, add mineral oil to the **ADH, LDC, ODC, H₂S,** and **URE** microtubes.

6. Cover the plastic tray and inoculated strip with the plastic cover, label, and place carefully in the incubation tray.

Results

Materials List per Table/Work Station

Potassium hydroxide, α-naphthol

Kovacs' reagent

Ferric chloride, 10% Solution

After incubation, the reactions of all three rapid identification methods must be read and interpreted, and the information can be used to identify the unknown. The actual process of identification will be the subject of the next exercise. Before the process of identification begins, all reactions must first be read and interpreted. Refer to the previous descriptions of each rapid identification test for a guide to interpreting results. Additional information in the form of pamphlets, guidebooks, and package inserts will be available to aid in interpreting results.

With the use of the manufacturer's package inserts, pamphlets, worksheets, and lab manual references, determine which reactions are positive or negative. Add the following reagents to these systems:

Enterotube™ II and Oxi/Ferm™ II Tube: Add several drops of Kovac's reagent to the appropriate chamber once all the other reactions are read. If this reagent leaks into adjacent chambers, colors may change, thus affecting the accuracy of your readings,

OR

Do not add any Kovac's reagent to either of these chambers. Rely on the indole test done on the API® 20 E System. This way you can review and perhaps reinterpret ambiguous results at a later time. (See the Appendix for a summary of results.)

API® 20 E System: Add 1 drop of potassium hydroxide (KOH) and 1 drop of α-naphthol and wait 10 minutes. If an obviously pink or red color develops, acetoin is present. If there is no color change or a slightly pink color develops, the reaction is negative.

Add 1 drop of ferric chloride to the TDA microtube and determine whether the color changes to dark brown, indicating the presence of pyruvate.

Once all reagents are added and other results interpreted, add 1 drop of Kovac's reagent (fume hood) to the "IND" microtube. Wait 2 minutes. A red color is positive for indole production. (See the Appendix for a summary of these results.)

Inventory

At the end of this laboratory, each person will have the following ready for incubation:

One Enterotube™ II Tube
One Oxi/Ferm™ Tube II
One API® 20 E System

NAME _____ DATE _____ SECTION _____

QUESTIONS

1. Which biochemical tests do all three rapid identification systems used in this exercise have in common?

2. What is the prime purpose of the Enterotube™ II Tube?

3. What is the prime purpose of the Oxi/Ferm™ Tube II?

4. Why does oil have to be added to some of the cupules of the API® 20 E System?

MATCHING

a. Enterotube™ II

b. Oxi/Ferm™ Tube II

c. API® 20 E System

_____ used to rapidly identify a suspected *Non-Enterobacteriaceae*

_____ tests for all three decarboxylase reactions

_____ used to rapidly identify a suspected Gram negative rod that ferments glucose

_____ test procedure that requires the microbe be mixed in a sterile saline solution

_____ rapid identification system that does not allow bacterial growth

_____ determines whether glucose is fermented or oxidized

MULTIPLE CHOICE

1. A Gram negative rod was isolated in the lab. A way to *quickly* presumptively identify it as either an *Enterobacteriaceae* or a *Non-Enterobacteriaceae* is through the:

   a. Oxi/Ferm™ Tube II   b. Enterotube™ II   c. IMViC   d. oxidase test

2. Microtubes filled with a saline suspension of a suspected Gram negative microbe are typical of:

   a. Kirby-Bauer   b. API® 20 E System   c. Oxi/Ferm™ Tube II   d. Enterotube™ II

3. The phenylalanine deaminase test is positive when which of the following is present?

   a. pyruvate   b. acetoin   c. dulcitol   d. oxidase

4. Glucose fermentation, urease, citrate utilization, and lysine decarboxylase are all tested for in which of the following?

   a. Oxi/Ferm™ Tube II   b. Enterotube™ II   c. API® 20 E System   d. all of these

5. Which aspect of the IMViC set of reactions is not tested for with the three rapid identification procedures done in this lab?

   a. indole   b. methyl red   c. Voges-Proskauer   d. citrate

6. The Oxi/Ferm™ Tube II utilizes a green medium for many of its carbohydrate tests. What other medium does this?

   a. dulcitol   b. O.F. basal medium

   c. Eosin-Methylene Blue agar   d. Enterotube™ II Tube

WORKING DEFINITIONS AND TERMS

API® 20 E System A rapid identification system made up of 20 microtubules utilized to identify both *Enterobacteriaceae* and many *Non-Enterobacteriaceae.*

Enterotube™ II A rapid identification system made up of 12 compartments of different media used to primarily identify suspected *Enterobacteriaceae.*

Oxi/Ferm™ Tube II A rapid identification system consisting of 12 compartments of different media used primarily to identify suspected *Non-Enterobacteriaceae.*

Identification of a Bacterial Unknown

Objectives

After completing this laboratory, you should be able to:

1. Confirm whether your unknown is either an *Enterobacteriaceae* or a *Non-Enterobacteriaceae*.

2. State the principle behind using a comparison chart, a flowchart, and a computer for identifying a bacterial unknown.

REVIEW

Observe and interpret the results of the three rapid identification systems inoculated in the previous exercise. Compare the results with similar tests done using the more traditional methods (e.g., citrate, H_2S, glucose fermentation, etc.) if these previous tests were performed with isolated colonies.

Various methods can be used to identify a bacterium once it has been isolated and grown in a pure culture. Serology is useful if a specific microbe is suspected, but it is considerably less effective if one of many different etiologic agents are capable of causing the disease in question. Biochemical reactions, therefore, remain the major method used to identify microbial pathogens.

You now have the results of over 20 different biochemical reactions to utilize in identifying your unknown. (Although nearly 50 different tests were performed in the previous labs, many of the individual tests are duplicated from one system to another.) The problem now is to determine which tests to use and how to evaluate these tests so that you have a reasonably good chance of identifying your unknown.

COMPARISON CHART

If you manage to narrow the choice of your unknown to only a few microbes, a *comparison chart* will be all that is necessary to distinguish between them. An example of such a chart is the use of the IMViC set of reactions previously covered. If used only to distinguish between *E. coli* and *Enterobacter aerogenes,* the test is usually reliable. If *Proteus vulgaris* and *Serratia marcescens* are added to the choices, the results become ambiguous and the additional tests must be considered before a final identification is made. A further complication is that not all strains of each genus and species are 100% positive or 100% negative for a particular reaction. That is, sometimes a specific microbe may have gained or lost a gene for a specific biochemical reaction and may not give the reaction that a particular comparison chart states it should give. For this reason, many comparison charts give not only a positive or a negative reaction for a particular biochemical reaction, but also the percentage of how many samples taken are positive or negative. (See Table 21.1.) See the Appendix for more detailed examples of comparison charts.

COMPARISON CHART FOR THE IMVIC TEST

Microbe	Indole	Methyl Red	Voges-Proskauer	Citrate
E. coli	89%(+)	99.9%(+)	100%(−)	100%(−)
Enterobacter aerogenes	100%(−)	98.4%(−)	85%(+)	82%(+)
Serratia marcescens	99%(−)	81.5%(−)	70%(+)	96%(+)
Proteus vulgaris	92%(+)	93%(+)	100%(−)	88%(−)

FLOWCHART UTILIZING LINE DIAGRAMS

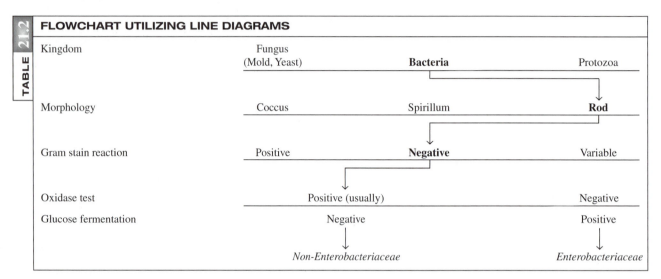

Kingdom	Fungus (Mold, Yeast)	**Bacteria**	Protozoa
Morphology	Coccus	Spirillum	**Rod**
Gram stain reaction	Positive	**Negative**	Variable
Oxidase test	Positive (usually)		Negative
Glucose fermentation	Negative		Positive
	Non-Enterobacteriaceae		*Enterobacteriaceae*

FLOWCHART FOR THE IDENTIFICATION OF ENTEROBACTERIACEAE

MICROBE: Gram-negative rod, glucose fermenting, oxidase negative.

Group I:
Phenylalanine Deaminase (PA or PAD) (+)   Possible microbes:
   or   *Proteus*
Tryptophane Deaminase (TDA)(+)   *Providencia*, or *Morganella*

(On the basis of this one test, all but three genera of *Enterobacteriaceae* have been eliminated from consideration. If this test is negative, skip to Group II.)

Urease (+) → *Providencia stuartii or alcalifaciens*

Inositol reaction (Carbohydrate)
   Inositol (+) = *Providencia stuartii*
   Inositol (−) = *Providencia alcalifaciens*

(On the basis of a negative urease test, the group of bacteria to be indentified has now been reduced to two. The inositol reaction will differentiate between the two finalists.)

Urease (−) → *Proteus, Providencia rettgeri, Providencia stuartii*

Microbe	Ornithine	H₂S	Indole	Rhamnose
Pr. mirabilis	+	+	−	
Pr. vulgaris	−	+	+	
Morganella morganii	+	−	+	
Prov. rettgeri	−	−	+	+
Prov. stuartii	−	−	+	−

Group II   Possible microbes:
Phenylalanine (−)   *Klebsiella*
Voges-Proskauer (+)   *Serratia*
                                                                          Enterobacter

FLOWCHART

A comparison chart is an effective way of distinguishing between a few different microbes, but when the number of choices increases, the number of comparisons necessary to identify a specific microbe also increases. Coupled with the fact that there are always a few reactions that come out "wrong" due to mutations and different subspecies, using a comparison chart to quickly identify large numbers of microbes becomes a lesson in frustration. A *flowchart,* however, is used to divide large numbers of microbial species into smaller, easier to manage numbers. Once the flowchart reduces the choices of large numbers of microbes to smaller groups, a comparison chart can then be used to make the final identification. The following flowchart is diagrammatic and would lead you to determine whether a microbe is an *Enterobacteriaceae* or a *Non-Enterobacteriaceae.* This chart is oversimplified, for the Kingdom is often determined by the patient's symptoms and both morphology and Gram stain reaction can be done in one step. (See Table 21.2.)

Another version of the flowchart does not utilize the line diagram seen above but still divides large numbers of microbes into smaller, easier to identify groups. This type of chart divides the microbes into major groups based on selected tests. An example of this type of flowchart, seen below, divides the family *Enterobacteriaceae* into six major groups. (See Table 21.3.) For a more systemic view of this flowchart, see the Appendix.

COMPUTER IDENTIFICATION

One of the more innovative developments of the last few decades in the field of medical microbiology has been the utilization of the computer to eliminate the problems of identifying the literally thousands of different biochemical variations of bacteria. By converting the different reactions found in the rapid identification tests to a series of numbers, these numbers can be entered into a preexisting computer program. The series of numbers that are generated depends on the type of rapid identification test performed.

The Enterotube™ II test develops a *five-digit number,* which is called an "ID Value." (See Fig. 21.1.)

ID VALUE WORKSHEET FOR ENTEROTUBE™ II

1. Review the reactions in the Enterotube™ II except for the Voges-Proskauer and indole production tests.

⚠ **CAUTION:** ONCE THE KOVAC'S REAGENT HAS BEEN ADDED, AND THIS REACTION HAS BEEN READ, THE TUBE SHOULD BE DISCARDED. IF YOU USE A DIFFERENT SOURCE FOR THE INDOLE REACTION, THAT IS, API® 20E, YOU CAN KEEP THE TUBE TO DOUBLE-CHECK YOUR RESULTS.

2. Circle the *number* underneath each positive reaction.

The *first* box located below the drawing of the glucose fermentation compartment will contain either zero, two, or three. "Zero" (0) indicates that the microbe does not ferment the glucose, two (2) indicates that glucose is fermented, and "three" (3) indicates that glucose is fermented with gas production. Since you cannot have gas production without glucose fermentation, the numeral "one" (1) is not one of the choices. If your unknown is

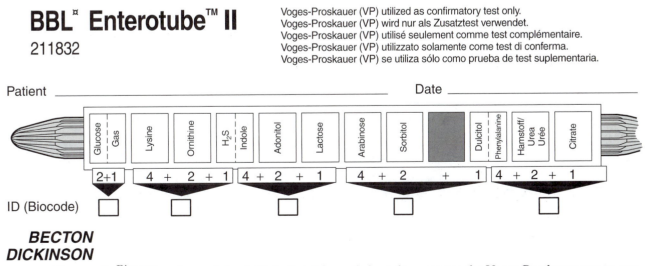

FIG. 21.1. Enterotube™ II Biocode worksheet. Note that this worksheet does not use the Voges-Proskauer test to generate the 5 digit I.D. value or Biocode.

not a glucose-fermenter and is oxidase postive, you should rely on the results produced by the Oxi/Ferm™ Tube II.

The *second* box will contain a number from zero through seven. "Zero" indicates that the microbe does not contain the enzyme for removing the carboxyl group from lysine or ornithine and does not produce hydrogen sulfide. "Seven" indicates that the microbe is positive for all three reactions.

The *third, fourth,* and *fifth* boxes will also contain numbers from zero through seven. Note that there is a blank space corresponding to the Voges-Proskauer test. The VP test is used as a confirmatory test and is not part of this particular ID Value number.

Note: There is an alternate ID Value number system that does use the Voges-Proskauer test as part of the primary identification. If this other system is used, the numbering system will be different, but the principle of acquiring these numbers is the same.

Once the five-digit ID Value is generated, it may be entered into a computer containing the appropriate software. The computer will give you the identification, or inform you that it is necessary to perform additional tests, or politely tell you to reevaluate your interpretation of the results because there is no such microbe. As stated above, such an approach takes into consideration the large variety of strains found within each species of bacterium. For example, *Escherichia coli* may have as an ID Value 20450, 20470, 22430, 24461, 24520, 24530, or 24560. If the ID Value 20070 is entered, the computer will inform you that you are dealing with either *E. coli* or *Salmonella paratyphi,* Type A. You will be told to perform additional or *confirmatory tests* to distinguish between these two microbes. In this case, a serological test

for *Salmonella* and a specific carbohydrate fermentation will be used for the final identification.

The *Oxi/Ferm™ Tube II* also develops a *five-digit number,* which is also called an ID Value. (See Fig. 21.2.)

The number is generated in much the same way as the Enterotube™ II except that the last digit includes the oxidase test. When this five-digit number is presented to the computer, you will be given the identity of the microbe, and additional or supplemental tests to perform, or be told to reevaluate your information. For example, the number 30317 indicates that the microbe in question is either *Pseudomonas aeruginosa* or *Pseudomonas putida.* You will then be told to inoculate a T-Soy slant and incubate at 42°C. The presence or absence of growth will then allow you to distinguish between these two microbes.

One of the most popular computer identification systems in use today is the *API® 20 E System.* This system utilizes a *seven-digit number,* which is generally considered superior to the five-digit systems just covered. (The more appropriate tests performed, the more accurate the results.) Note that the last of these seven digits include the results of the oxidase test (OX). In addition to this seven-digit "profile number," several other supplemental tests can be performed on this test strip, such as nitrate reduction and catalase. (See Fig. 21.3.)

The *first* oval of the "*Profile Number*" will contain any digit from zero through seven. A positive ONPG reaction would be "1," arginine would be "2," and lysine would be "4." Note that although a zero through seven digit is used, the order of the digits is different from the Enterotube™ II or the Oxi/Ferm™ Tube II. That is, these two tubes use a 4–2–1 sequence, whereas the API® 20E System utilizes a 1–2–4 progression.

BBL® Oxi/Ferm™ Tube II

4373254

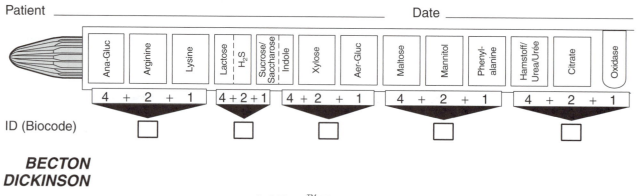

BECTON DICKINSON

FIG. 21.2. Oxi-Ferm™ ID Value worksheet.

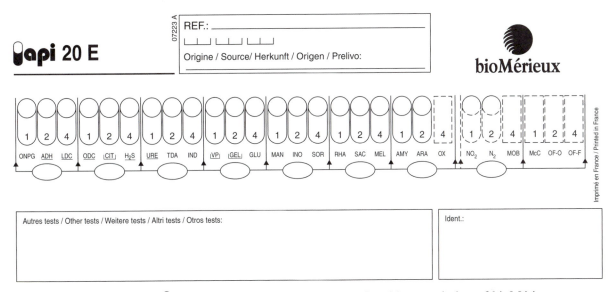

FIG. 21.3. API 20 E® "Profile number" Worksheet. Reprinted by permission of bioMérieux.

The remaining ovals of the API® 20E System follow the same 1–2–4 progression. The last component also reads for the oxidase test (for a total of 21 reactions).

The API® 20E System also allows for additional tests and thus "Additional Digits" to be generated in case of ambiguous results.

As with the other tests, once the seven-digit Profile Number is generated, it may be entered into a computer containing the appropriate software for identification. The advantage of this particular system is that the tests are flexible enough to identify members of both *Enterobacteriaceae* and *Non-Enterobacteriaceae*. As with the other two computer programs, confirmatory tests are listed whenever the Profile Number produces ambiguous results. For example, the Profile Number 1144452

indicates that the microbe is *E. coli,* whereas 1144500 indicates that the microbe is either a *Shigella* species capable of causing a severe intestinal infection, or *E. coli.* The program will list which confirmatory tests to perform so a final identification can be made. In addition to this extra information, the program will tell you if you have isolated extremely rare or dangerous microbes.

Results

Determine which method to use to identify your unknown(s): comparison chart, flowchart, or computer program. When the unknown is determined, report the results to your laboratory instructor.

NAME _____ DATE _____ SECTION _____

QUESTIONS

1. How would you determine whether a Gram negative rod can be classified as belonging to *Enterobacteriaceae* or a *Non-Enterobacteriaceae* using these three rapid identification tests?

2. Which of the three rapid identification tests can be used to identify members of both groups of Gram negative rods?

MATCHING

a. *Enterobacteriaceae*

b. *Non-Enterobacteriaceae*

c. *Proteus mirabilis* (see Table 21.3)

d. *Proteus morganii* (see Table 21.3)

e. Enterotube™ II

f. API® 20 E System

g. flowchart

h. Oxi/Ferm™ Tube II

_____ methodically reduces choices of microbes from a large number to increasingly smaller numbers

_____ orthinine (+), H$_2$S (−), indole (+)

_____ Gram negative rod, glucose fermentation positive, oxidase negative

_____ ornithine (+), H$_2$S (+), indole (−)

_____ Gram negative rod, glucose oxidation positive, oxidase positive

_____ tests for both *Enterobacteriacea* and many *Non-Enterobacteriaceae*

_____ rapid identification procedure designed primarily for *Enterobacteriaceae*

MULTIPLE CHOICE

1. You have isolated a Gram negative rod. You have only an Enterotube™ II and an Oxi/Ferm™ Tube II to work with but, due to budget cuts, can only inoculate one. Which test would most likely determine which tube to inoculate?

   a. nitrate reduction   b. oxidase   c. catalase   d. coagulase

2. You have inoculated and grown your unknown in one (or all) of the rapid identification systems used in this lab. Which is the first reading you would take in order to determine whether you have an *Enterobacteriaceae*?

   a. ONPG   b. phenylalanine deaminase or TDA   c. glucose fermentation   d. citrate

3. You are reading the results of the Enterotube™ II ID value worksheet (see Fig. 21.1). Which number is impossible to have?

   a. 12441   b. 32441   c. 22441   d. 02441

4. You are reading an Oxi/Ferm™ Tube II and develop an ID value (see Fig. 21.2). Which number indicates that you inoculated it with the "wrong" microbe for this system to identify?

   a. 01121   b. 01123   c. 01125   d. 41120

5. After inoculation with a rapid ID system you decide you would like to do a catalase and nitrate reduction (to nitrate) test. Which of the following would allow you to perform these additional tests?

   a. API® 20 E System   b. Oxi/Ferm™ Tube II   c. Enterotube™ II

V FOOD AND ENVIRONMENTAL MICROBIOLOGY

B acteria, viruses, fungi, and protozoan cysts are common contaminants of many food products that we consume regularly. Milk and water used in the preparation of juice from frozen concentrates, and ice tea beverages are often contaminated with microbes. Meat products such as poultry, beef, and pork have also been found to contain several different types of organisms. The degree of contamination that may be found in the food product depends on several factors, including the environment in which the food was prepared; the initial number of contaminating microbes found in the food prior to its processing; the degree of sanitation used during food processing procedures; and the manner in which the food was packaged or stored prior to purchase.

Our environment is also a source of microbes. Soil from all over the world is routinely sampled for organisms that provide us with antibiotics and other drugs used to treat human maladies. The relationships between these microbes and between them and us are also a source of interest; microbial ecology is constantly being studied in its relationship to the environment and to humans.

Identification and Quantitation of Microbial Numbers in a Water Sample

Objectives

After completing this lab, you should be able to:

1. Distinguish between a presumptive test and a confirmation test for the identification and quantitation of coliform bacteria in water.

2. Biochemically distinguish between *Escherichia coli* and *Enterobacter aerogenes*, both of which are possible water contaminants.

Fecal contamination of water can serve as a source of primary pollution. If the contaminated water is used for food processing, potential infection or disease can be transmitted either with consumption of the contaminated water or the foods themselves processed with it. Several organisms found in polluted water can make you quite ill. These organisms include certain strains of *E. coli* and *Enterobacter aerogenes*, *Klebsiella pneumoniae*, *Salmonella species*, *Shigella dysenteriae*, *Vibrio cholera*, Hepatitis A, and the protozoans, *Entamoeba histolytica* and *Giardia lamblia*.

U.S. Public Health Service agencies are responsible for the continuous screening of water supplies to reduce the threat of ingestion of waterborne pathogens, which can lead to serious infection or disease. The presence of *E. coli* in water, a common inhabitant of the intestine, is an indication of fecal contamination of water. Microbiologists determine the *Coliform count* (usually the *E. coli* number) to determine the quantity of this intestinal bacterium in the water. Coliforms are Gram negative facultative anaerobic rods. They ferment lactose and produce acid and gas from it as end products of metabolism.

The Coliform count procedure requires that a sample of water be tested first to determine whether bacteria are present. A *presumptive test* determines the presence of Coliforms by demonstrating that the growth and gas production occurred after inoculation of the water sample into a lactose fermentation tube. The Durham inner tube shows the evidence of gas production, and the red/orange to yellow color conversion of the phenol red indicator confirms that acids are produced by the bacterial metabolism of lactose. See Exercise 11 for a review of the Durham fermentation tube.

A *confirmation test* is then performed on the sample in order to further verify the Coliform contamination of the water sample. The test utilizes differential and/or selective laboratory media. A tube that shows a positive reaction in the presumptive test is used as the sample inoculum for this test. Eosin-Methylene Blue (EMB) or MacConkey's agar (MAC) are inoculated in the confirmation test from a loopful of a positive reaction observed in the presumptive test. Both of these solid agar media favor the growth of Gram negative rods and can determine whether the organism is a lactose fermenter or non-lactose fermenter. On EMB agar, lactose fermenters have darkly pigmented colonies, whereas the non-lactose fermenters produce clear to colorless colonies. Very often *E. coli* produces a flat colony with greenish metallic sheen. On MAC agar, lactose fermenters form reddish-purple colonies, whereas non-lactose fermenters form colorless colonies. See Exercise 13 for a review of these differential growth media.

A nutrient agar slant is also prepared from the suspected Coliform colony from EMB or MAC agar, Gram stained to verify that the organism is a Gram negative rod, and then it is inoculated into the IMViC set of LAB media. The IMViC tests verify the presence of *E. coli* organisms specifically over other types of possible enteric pathogens, which can be found in a contaminated water specimen. (*Salmonella* and *Shigella* species traditionally are non-lactose fermenters.)

⚠️ (REMEMBER: IMVIC MEANS INDOLE, METHYL RED, VOGES-PROSKAUER, AND CITRATE TEST REACTIONS. SEE EXERCISES 11 AND 19 FOR A REVIEW OF THESE REACTIONS.)

PRESUMPTIVE TEST: ANALYSIS OF A CONTAMINATED WATER SAMPLE FOR THE PRESENCE OF COLIFORMS

This activity will detect the presence of *Escherichia coli* and/or *Enterobacter aerogenes* in a contaminated unknown water sample. Since both organisms are lactose fermenters and ferment glucose in order to produce acid and gas products, it is necessary that the Coliform count procedure be modified by inoculating the sample into a Simmons citrate tube. *E. coli* does not utilize citrate as a sole carbon source; *E. aerogenes* does utilize citrate. See Table 22.1 for the Biochemical Test reactions that distinguish both organisms.

Materials List per Table/Work Station

Broth Cultures of *E. coli* and *E. aerogenes*

Unknown water sample of either one or a mixture of both organisms

Three glucose fermentation tubes

Three lactose fermentation tubes

Three Simmons citrate slant tubes

Three T-Soy broth tubes and Kovac's reagent

Six Methyl Red—Voges-Proskauer (MR-VP) broth tubes and reagents

Three plates of MacConkey agar

3% hydrogen peroxide

Oxidase reagent

PROCEDURE

1. Inoculate your unknown water sample and both known broth cultures each into a separate tube of Simmons citrate medium.

2. Streak three plates of MacConkey agar: one with your unknown water sample and one with each of the two broth cultures. Streak the plates for isolated colonies. Also inoculate a glucose fermentation tube, lactose fermentation tube, T-Soy broth tube, and MR-VP broth tubes (remember to inoculate two MR-VP broth tubes) with each sample. Incubate all tubes and plates at 37° C for 24 hours.

Results

1. Observe and record the results of the inoculations of your unknown water sample and both known organisms in Table 22.2.

2. Perform a Gram stain on your unknown sample. If the MacConkey plate shows the presence of two different colonies, perform a Gram stain on each one. Also record the results in Table 22.2.

TABLE 22.1 KNOWN REACTIONS FOR BACTERIA[a]

Laboratory Media/Tests	E. coli	E. aerogenes
1. Glucose fermentation	Acid + gas	Acid + gas
2. Lactose fermentation	+ (or −)	+
3. Citrate utilization	−	+
4. Indole production from tryptophan	+	−
5. Methyl red test	+	−
6. Voges-Proskauer test	−	+
7. Catalase activity	+	+
8. Oxidase activity	−	−

[a] From the shorter Bergey's *Manual of Determinative Bacteriology*, 8th ed. (Baltimore, Md.: Williams and Wilkins Co., 1977), p. 101.

TABLE 22.2 BACTERIAL REACTIONS OBSERVED WITH VARIOUS LABORATORY MEDIUM AND METHODS

Laboratory Media/Tests	Positive or Negative Result		
	Unknown Sample	E. coli	E. aerogenes
Glucose fermentation tube			
Lactose fermentation tube			
Citrate utilization			
Indole test			
Methyl red test			
Voges-Proskauer test			
MacConkey agar growth			
Gram stain			

3. Compare the results of your two control cultures, *E. coli* and *E. aerogenes* with Table 22.1 and with your unknown. Determine whether your unknown is *E. coli, E. aerogenes,* or a mixture of both.

QUANTITATION OF MICROBIAL NUMBER IN A WATER SAMPLE

The quantitation of microbial numbers in a water sample is determined by the *Standard Plate Count Method.* Diluted samples to be tested for organism numbers are mixed with standard quantities of melted and partially cooled agar. After 48 hours of incubation at 37° C, visible colonies are counted with the aid of a colony counter. The total number of colonies counted multiplied by the reciprocal of the dilution made of the sample determines the standard plate count per milliliter of sample. For example, if you counted 52 colonies and your dilution of the unknown sample was 1:100, $100 \times 52 = 5200$ bacteria/ml in the original sample. (Accurate quantitative measurements require a range of colony counts between 30 and 300 colonies/plate.)

Materials List per Table/Work Station

Three test tubes containing 9 ml of sterile water

Six melted Tryptone Yeast agar tubes

Six sterile Petri dishes

PROCEDURE

1. Use your contaminated unknown water sample and prepare the following dilutions:

   a. Select three tubes containing 9 ml of sterile water each.

⚠ **CAUTION: NEVER PIPETTE BY MOUTH; ALWAYS USE A BULB!**

   b. Aseptically transfer 1.0 ml of the contaminated water sample to tube number one, mix well, and transfer 1 ml of this mixture to tube number two. Again, mix tube number two and aseptically transfer 1 ml from tube number two to tube number three. (You can utilize the same pipette for the preparation of these serial dilutions.) You have now prepared a 1:10, 1:100, 1:1,000 dilution of the contaminated water sample. (See Fig. 22.1.)

⚠ **REMEMBER TO MIX THE SAMPLES WELL AT EACH DILUTION STEP.**

2. Add 1 ml of each dilution made to the base of each of two empty Petri dishes. Label each set of two plates with dilutions made: 1:10, 1:100, 1:1,000. To each Petri dish containing a different dilution of water sample, pour 10 ml of melted tryptone glucose yeast agar. Rock and rotate the contents of each plate to distribute the mixture and allow the agar to solidify.

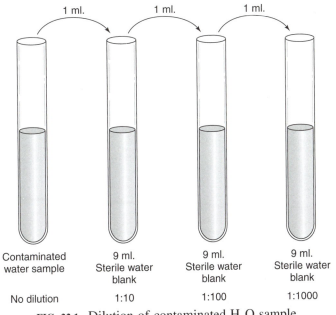

| 1 ml. | 1 ml. | 1 ml. |

Contaminated water sample

No dilution

9 ml. Sterile water blank

1:10

9 ml. Sterile water blank

1:100

9 ml. Sterile water blank

1:1000

FIG. 22.1. Dilution of contaminated H_2O sample.

Note: Make sure the agar is warm to the touch before you pour it into the sterile Petri plate containing the diluted water sample. If it is too hot, it will destroy the organisms present in the diluted sample dispensed on the plate.

3. Invert the agar plates and incubate them for 48 hours at 37° C.

Results

Count the colonies on all plates and calculate the number of bacteria/ml in the original sample. You must have 30 to 300 colonies on the plate for an accurate plate count. Be sure to take into account your dilution factor when calculating your bacterial number. Fill in Table 22.3

The average number of bacteria/ml in the unknown water sample was determined to be: _____.

TABLE 22.3 COLONY COUNTS AND DETERMINATION OF BACTERIAL NUMBER IN THE ORIGINAL SAMPLE

Colony Count	Dilution	Number of Bacteria/ml
	1:10	
	1:10	
	1:100	
	1:100	
	1:1,000	
	1:1,000	

NAME _____ DATE _____ SECTION _____

QUESTIONS

1. How would you distinguish between *E. coli* and *Enterobacter aerogenes* if they were present together in a mixed culture? Identify the media you would select and results of the tests you would use to differentiate between these organisms.

2. Why is it not necessary to perform a Gram strain on the microbes isolated from MacConkey or EMB agar used in the presumptive identification of Coliforms?

3. Could MacConkey agar be used instead of Tryptone Glucose Yeast agar to determine microbial numbers from a contaminated water sample? Explain.

4. What differences exist between *E. coli* and *Enterobacter aerogenes* regarding their IMViC reactions?

MATCHING

TEST	CHEMICAL REACTION
a. citrate	_____ fizzing occurs upon addition of 3% H_2O_2
b. oxidase	_____ turns red when mixed with Kovac's reagent
c. Voges-Proskauer	_____ red-purple colonies on MacConkey agar
d. gas production	_____ blue-colored slant from original green-colored slant tube
e. catalase	_____ acetomethyl carbinol
f. lactose fermentation	_____ a purple color occurs on the paper towel within 30 seconds after adding the reagent
g. indole	_____ bubble seen in glucose fermentation Durham tube

MULTIPLE CHOICE

1. In a presumptive test reaction, Coliforms inoculated into a lactose fermentation tube produce _____ and _____.

   a. acid, gas   b. oxidation, fermentation   c. acetoin, indole   d. pyruvate, indole

2. When *E. coli* grows on a MacConkey agar plate, its colony is what color?

   a. green   b. black   c. pink   d. colorless

3. Indole is identified in a T-Soy broth tube following the addition of a reagent called:

   a. potassium hydroxide   b. Kovac's reagent   c. oxidase reagent   d. methyl red

4. The Voges-Proskauer test is designed to detect this metabolic product.

   a. acetoin   b. indole   c. oxidase   d. acid conditions

5. Name the indicator used in the preparation of either glucose or lactose fermentation tubes.

   a. methyl red   b. indole   c. alpha naphthol   d. phenol red

6. The catalase enzyme is detected by the addition of 3% hydrogen peroxide to a colony. What happens if the colony is catalase positive?

   a. the colony bubbles or fizzes
   b. the colony turns purple
   c. the colony dissolves
   d. coagulation takes place

7. You determined an average colony count of 30 colonies using a 1:1000 dilution of the original sample. What is the average number of bacteria in the original sample?

   a. 30   b. 300   c. 3000   d. 30,000

WORKING DEFINITIONS AND TERMS

Coliform A common name for intestinal bacteria, most often *E. coli,* that is used as an indicator of fecal contamination of water. The number of bacterial cells per ml of water is termed the Coliform count.

Confirmation test The second of two tests measuring the presence of Coliform bacteria in a sample of water by using differential growth media such as MacConkey or EMB.

Presumptive test An initial test to determine whether a water sample is fecally contaminated. If this test comes out positive for lactose fermentation, a second test is performed to confirm lactose utilization (confirmation test).

Identification of Microbes in Beef and Poultry and the Quantitation of Microbial Numbers

EXERCISE 23

Objectives

After completing this lab, you should be able to:

1. Explain the importance of refrigerating food until it is ready to be cooked or eaten.

2. Distinguish between *E. coli* and *Salmonella* species by way of cultural morphology and biochemical tests.

In general, the public's health is protected by those rules and regulations established by food manufacturers and approved by the Food and Drug Administration with the aim of keeping potential food-borne diseases within a population to a minimum. Foods that are hazards to a person's health are those that:

1. Possess large numbers of bacteria that can cause an *infection* by mere ingestion of the food, or

2. Have microbial products of metabolism, which causes *intoxication* of the body.

To prevent food contamination with bacteria, preservatives are added to many foods to extend their product shelf life. They are then packaged, frozen, and dated so that they are consumed within a designated time period before the threat of spoilage is possible. Sometimes, however, the food handling is faulty and persons do get sick from food-borne microbes. The generalized symptoms of food poisoning are diarrhea, vomiting, and abdominal pain or cramps.

Recently, increasing numbers of emerging infections have occurred worldwide in fresh meat products such as chopped meat where *E. coli* has been identified and in poultry where *Salmonella* species have been recognized. Although high standards of sanitation and hygiene are maintained in food preparation, food-borne bacterial contaminants are becoming a threat to the

general public health. Consumers are therefore being warned to thoroughly cook all beef and poultry products before eating.*

Microbes have also been identified in cooked foods that are often eaten cold, such as processed meats, custards, and cheese. Diarrhea, a common problem with these infections, is caused by organisms such as *Bacillus cereus,* which contaminates meats, vegetables, soups, stews, sausage, sauces, and desserts; *Listeria monocytogenes,* which contaminates soft cheeses and paté, and *Shigella* species, which taint prepared potatoes, chicken, tuna, and shrimp salads.

In this laboratory exercise, we will determine the presence of organisms in chopped meat and chicken purchased for family consumption, and we will also describe the correct handling of these products. We will also determine the approximate number of microbes present in the specimen.

---

*The pathogenic *E. coli* strain 0157:H7 in undercooked hamburger meat requires that all fast food restaurants cook beef products well. Similarly, the threat of *Salmonella* infection requires that eggs and poultry be well cooked before consumption. These food samples may contain virulent strains of microbes as opposed to the avirulent strains available in the laboratory.

MICROBIAL PRESENCE IN A FOOD PRODUCT AND COLONY COUNTS

You will be provided with 10^{-1} dilutions of a suspension of chopped meat and of poultry. These suspensions were prepared for use in this laboratory by placing 20 grams of animal tissue in 180 ml of sterile water and mixing them in a blender for 5 minutes. There are two preparations of beef and two of poultry. One 10^{-1} dilution of beef is labeled *normal handling*. The other 10^{-1} dilution of beef is a preparation made from beef that is allowed to sit on a counter overnight at room temperature before processing. It is labeled *abnormal handling*.

The same procedure is followed for poultry. One 10^{-1} dilution of specimen is labeled *normal handling* and the other is called *abnormal handling*.

Materials List per Table/Work Station

Suspensions of chopped meat, normal and abnormal

Suspension of poultry, normal and abnormal

12 sterile Petri dishes

12 capped sterile test tubes

12 tubes containing 9 ml of sterile distilled water

1 ml pipettes with dispensers

12 tubes of 20 ml melted MacConkey agar

4 broth tubes of T-Soy medium

8 broth tubes of MR-VP

4 slant tubes of Simmons citrate medium

4 deeps of Sulfide-Indole-Motility media

PROCEDURE

1. Obtain a 1 ml suspension of the 10^{-1} dilutions of normal and abnormal beef preparation and 1 ml suspension of the 10^{-1} dilutions of normal and abnormal poultry preparation.

2. A total of 12 tubes containing 9 ml of sterile water should be arranged so that 3 tubes are in sequence with each 1 ml suspension of specimen. Mix 1 ml of the normal handling beef suspension into the 9 ml of sterile water. You now have a 10^{-2} dilution. Remove 1 ml of the 10^{-2} dilution that has been mixed and transfer it into another 9 ml tube of sterile water. The second tube is mixed well to give a 10^{-3} dilution. Transfer 1 ml of this mixture into a third 9 ml of sterile water to prepare a 10^{-4} dilution. (See Fig. 23.1.) Repeat the above procedure (step 2) for the other three suspensions.

3. Label the bottom of three Petri dishes: normal beef 10^{-2}, normal beef 10^{-3}, and normal beef 10^{-4}. Transfer 1 ml of each normal beef dilution to the base of these sterile Petri dishes. Follow the same procedure for the abnormal beef, normal poultry, and abnormal poultry. You should now have 3 plates of each dilution for each specimen. A total of 12 plates should be used.

4. Pour 20 ml of melted but cooled MacConkey agar over each dilution dispersed into the base of a sterile Petri plate and cover the plate with its lid. (The agar should be melted but not steaming hot; it should be cool to the touch.) Swirl the plate on the lab bench several times to disperse the microbes that may be in the specimen.

5. Allow the agar to harden for about 10 minutes. Invert the plates and incubate them at 30° C for 48 hours.

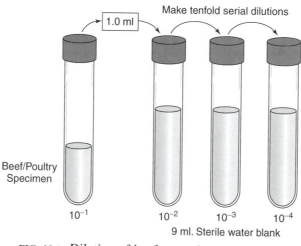

FIG. 23.1. Dilution of beef or poultry suspension.

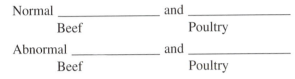

TABLE 23.1	COLONY COUNTS OF BEEF & POULTRY SPECIMEN'S				
		Beef Specimens		Poultry Specimens	
	Dilution	Normal	Abnormal	Normal	Abnormal
	10^{-2}				
	10^{-3}				
	10^{-4}				

What is the average number of bacteria ml of beef and poultry specimen?

Normal _____ and _____
            Beef   Poultry

Abnormal _____ and _____
             Beef   Poultry

Results

After incubation, observe the plates, record your results in Table 23.1 (according to steps e and f below), and answer the following questions:

a. Are microbes found in the beef and poultry specimens?

b. Which dilution of the specimen has countable colonies? (A total of 30 to 300 colonies/plate is countable.)

c. Do you see lactose-fermenting colonies (pink to lavender) and non-lactose-fermenting colonies (colorless) on the agar?

d. Which plates had more growth/dilution—the normal or abnormal manner of handling specimens?

e. Count any *pink* colonies observed on the plates from the *beef* suspension and record your results in the table below.

f. Count any colorless colonies observed on the plates from the *poultry* suspension. Also record your results in the table below.

RECOGNITION OF ORGANISM GENERA

Determine whether the colonies isolated in the above experiment are *E. coli* or *Salmonella* species.

1. Select a pink colony isolated from a MacConkey agar plate of the beef suspension. Inoculate it into the IMViC test reaction tubes. Perform the same inoculations with a colorless colony isolated from the MacConkey agar plate of the poultry suspension. Incubate all media for 48 hours at 37° C.

2. Refer to Table 23.2 for the reactions that identify *E. coli* and *Salmonella* species and record your results.

TABLE 23.2	COMPARISON OF E. COLI & SALMONELLA			SUSPECTED COLONIES	
	Media and Reaction	E. coli	Salmonella	E. coli	Salmonella
	(MacConkey)—lactose fermenter	+	−	Pink Colony	Colorless
	(Tryptic nitrate)—indole production	+	−		
	(MRVP)—methyl red + or −	+	+		
	(MRVP)—Voges-Proskauer-Acetoin	−	−		
	(Citrate)—citrate utilization	−	+		
	Sulfide-Indole-Motility—H_2S (+ or −)	−	+		

Source: Shorter Bergey's Manual of Determinative Bacteriology.

NAME _____ DATE _____ SECTION _____

QUESTIONS

1. If you wanted to determine whether *E. coli* or *Salmonella* species were motile, what media would you select for this purpose? What would be your findings for each organism, and how would you interpret or read the results in the tube inoculated?

2. What genera of organism are commonly isolated from poultry products and can lead to food-borne illness following improper preparation of that food? How do you best prevent this form of food-borne illness from occurring?

3. What tests in the IMViC series are unable to distinguish between the genus *Escherichia* and the genus *Salmonella*? Explain.

4. Food contaminated with *Staphylococcus* is also a major problem in the catering industry. Explain why the procedures utilized in this exercise cannot be employed to detect and quantify this particular microbe.

MATCHING

GENUS OF ORGANISM

a. *Eschericha*

b. *Salmonella*

CHEMICAL REACTION

_____ methyl red positive

_____ acetoin negative

_____ citrate utilization positive

_____ H₂S positive

_____ lactose utilization positive

_____ indole negative

MULTIPLE CHOICE

1. When H_2S is produced in a SIM tube, the tube will appear:

   a. blue   b. yellow   c. black   d. green

2. When methyl red is added to a culture tube of bacterial growth and the tube remains straw colored, this means:

   a. acid was produced   b. glucose was fermented   c. acetoin was not utilized   d. none of these

3. A tryptophanase acts on tryptophan to produce:

   a. acetone   b. indole   c. iron salt   d. catalase

4. At which dilution would you expect to find the lowest microbial numbers?

   a. 10^{-1} b. 10^{-2} c. 10^{-3}

WORKING DEFINITIONS AND TERMS

Avirulent Strain or form of a microbe has few, if any, characteristics that make it dangerous (e.g., does not form toxins or have a protective capsule).

Virulent Strain or form of a microbe has characteristics that make it more dangerous (e.g., resistant to numerous antimicrobial drugs, produce toxins, etc.).

EXERCISE

24 Soil Microbiology

Objectives

After completing this lab, you should be able to:

1. Describe the nature of soil microorganisms.

2. Recover microorganisms from soil.

Soil is not an inert substance. It contains both inorganic and organic materials; in fact, the top layer of soil is teeming with microscopic and macroscopic organisms. The inorganic components consist of rocks, minerals, water, and gases, such as carbon dioxide, oxygen, and nitrogen. The most abundant inorganic materials are pulverized rock and the minerals and elements that *weathering* (mechanical breakdown) of the rock releases into the soil. The most abundant elements in most soils are silicon, aluminum, and iron. The organic components consist of *humus* (nonliving organic matter) and living organisms.

All the major groups of microorganisms (bacteria, fungi, algae, protists, and viruses) are present in soil. *Topsoil,* the surface layer of soil, contains the greatest number of microorganisms because it is well supplied with oxygen and nutrients. The most abundant microorganisms, constituting about 80% of the microbes present, are bacteria, especially aerobic bacteria. One of the most common of the soil bacteria is *Bacillus subtilis.* When the soil is well hydrated, the number of fungal and protozoan organisms increases. Soil viruses are predominantly *bacteriophages* (viruses that infect bacteria).

The most common human pathogens in soil are the spore-forming bacteria, *Bacillus* and *Clostridium.* One species of *Bacillus* is the causative agent of *an-* *thrax,* a fatal respiratory disease that infects cattle but can transfer to humans. The spores of a *Clostridium* species, when introduced into a deep puncture wound, germinate to cause the disease *tetanus.* Fungal diseases, called *mycoses,* can also be contracted by workers whose occupation (farming, construction) involves digging in soil. Pathogenic fungi associated with soil include *Blastomyces, Coccidioides, Aspergillus,* and *Histoplasma;* these organisms are associated with a deep systemic respiratory disease, characterized as PPI—*Primary Pulmonary Infection.*

THE RECOVERY OF MICROORGANISMS FROM SOIL

Materials List Per Table/Workstation

Sterile sample container/spatula/sieve/sterile saline/beaker/metric balance/stir plate/stir bar

1 ml and 10 ml pipettes

Empty sterile tube (10 ml)

Microscope slides/cover slips/microscopes

Bent glass rods (hockey sticks)

T-Soy Agar plates

Sabouraud Dextrose Agar plates

Sterile saline tubes (9 ml)

Magnifying glass/dissecting microscope

PROCEDURE

1. Obtain a soil sample from the environment. Use a sterile recovery container and a spatula. Obtain not just the surface soil, but dig several inches down to obtain deeper soil.

2. Sift the soil sample through a sieve to remove rocks, twigs, and so on.

3. Place about 100 ml of sterile saline in a beaker.

4. Weigh out about 5 grams of soil and suspend the soil in the saline in the beaker. (If results produce sparse numbers of soil microorganisms, the quantity of soil here may be increased.)

5. Place the beaker on a stir plate and stir for about 15 to 30 minutes. You are trying to dislodge microbes that adsorb to the soil particles.

6. While the soil suspension is still stirring, pipette 10 ml of the suspension into a sterile tube.

⚠ **REMEMBER: NEVER MOUTH PIPETTE!**

7. Optional: The suspension may now be centrifuged, but various microorganisms will be deposited and lost in the pellet.

8. This suspension will now be used to make microscopic observations; to inoculate various agars; and to make serial dilutions, if necessary.

   a. *Microscopic observation:* Mix the tube well by rolling it between the palms of your hands. Using a dropper, place a few drops of suspension on a microscope slide. Cover with a glass cover slip. Observe under low or high dry power of the microscope. You are specifically searching for the presence of protozoa (and algae, if the soil sample came from an area exposed to sunlight). Sketch your observations.

   b. *Agar inoculation:* Mix the tube well by rolling it between the palms of your hands. Using a 1 ml pipette, pipette 0.5 ml of suspension onto the surface of a Trypticase Soy Agar (TSA) plate. Repeat using a Sabouraud Dextrose Agar (SDA) plate. Spread the suspension on each plate using a sterile glass hockey stick. (See Exercise 15, Fig. 15.5.) Incubate both plates at room temperature for 48 to 96 hours.

   c. *Serial dilutions:* If the suspension plates from above produce growth that is too heavy, serial dilutions of the suspension may be made. Mix the

suspension tube well by rolling it between the palms of your hands.

Using a 1 ml pipette, pipette 1 ml of suspension to a tube containing 9 ml of sterile saline. Mix well. This is a tenfold dilution. Now, using a fresh pipette, pipette 1 ml of the tenfold dilution to a tube containing 9 ml of sterile saline. Mix well. This is a hundredfold dilution of the original suspension. By continuing this dilution procedure, you may make further dilutions. Use these dilution tubes to inoculate the agars, as above in step 8b.

RESULTS

a. *Microscopic observation:* Sketch any protozoa or algae observed:

b. *Agar inoculation:* The TSA plate will grow both bacteria and some fungi; the SDA plate is selective for fungi. Using a dissecting microscope or a magnifying glass, examine and sketch any mold observed:

Gram stain setup

Spore stain setup

PROCEDURE

1. Observe for the presence of bacterial colonies on the plates from step 8b and step 8c above.

⚠ **REMEMBER: DO NOT USE ANY PLATES THAT CONTAIN MOLD!**

2. Choose a bacterial colony, and, using a loop, streak it for isolation onto a fresh TSA plate. (See Exercise 2 for a review of how to make an isolation streak plate.) You may repeat this step a number of times using alternatively available colonies.

3. Incubate the fresh plates at room temperature for 48 hours.

Results

After incubation, make smears of any isolated pure colonies of soil bacteria. (See Exercise 2 to review how to make a smear preparation.) Gram stain the smears (See Exercise 5). Observe under oil immersion on the microscope. (In addition, the spore stain, Exercise 6, may also be employed to search for the presence of spores.) Sketch your observations:

Note: DO NOT OPEN ANY PETRI PLATES WITH MOLD! Mold spores can easily become air-borne and can cause infection, allergy, or laboratory contamination.

ISOLATION OF SOIL BACTERIA

Materials/Group List

Incubated agar plates from "The Recovery of Microorganisms" above

Fresh TSA plates

Inoculating loops

Microscope slides

Oil immersion

Lens paper

ISOLATION OF SOIL BACTERIOPHAGES OF *BACILLUS SUBTILIS*

Materials List Per Table/Workstation

Nutrient Agar plate

1 ml pipettes

24-hour broth culture of *B. subtilis*

Tube of soft overlay agar (0.7% agar), melted

Waterbath set at 45° C

PROCEDURE

1. Follow steps 1 through 7 from the first procedure above.

2. After centrifugation, decant the supernatant into a fresh sterile tube.

3. You must work quickly here: Obtain a tube of melted soft overlay agar from the waterbath. Wipe off all of the water from the surface of the tube. Using a 1 ml pipette, pipette 0.3 ml of a broth culture of *B. subtilis* into the soft agar tube.

4. Using a 1 ml pipette, aseptically transfer 0.5 ml of the supernatant into the soft agar tube. Mix the agar tube by rolling it between your hands. Do not allow the agar to solidify.

5. Immediately, aseptically pour the soft agar onto the surface of a Nutrient Agar plate. Replace the lid and without picking up the plate, rotate it gently in a 6 to 8-inch circle on the surface of the table to distribute the agar evenly.

6. Allow the soft agar to solidify.

7. Invert and incubate the plate at 37° C for 24 hours.

8. In your search for bacteriophages, you may repeat this procedure successively until all of the supernatant has been consumed.

Results

Observe for the presence of bacteriophage plaques (clear zones of lysis by a bacterial virus on the bacterial lawn).

NAME _____ DATE _____ SECTION _____

QUESTIONS

1. Describe the organic nature of soil.

2. Describe the inorganic nature of soil.

3. In which layer of soil are microorganisms most abundant?

4. Which type of microorganism is the most abundant in soil? What gases are found in soil?

5. What minerals are most abundant in soil?

6. Which human pathogens are associated with soil?

MATCHING

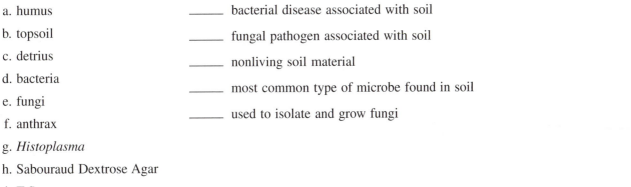

a. humus

b. topsoil

c. detrius

d. bacteria

e. fungi

f. anthrax

g. *Histoplasma*

h. Sabouraud Dextrose Agar

i. T-Soy agar

_____ bacterial disease associated with soil

_____ fungal pathogen associated with soil

_____ nonliving soil material

_____ most common type of microbe found in soil

_____ used to isolate and grow fungi

MULTIPLE CHOICE

1. Based on previous laboratory exercises and lecture material, which of the following would be expected to be recovered only from a deep soil sample?

    a. *Bacillus* bacteria   b. *Aspergillis* fungi   c. *Clostridium* bacteria   d. *Histoplasma* fungi

2. Which of the following soil microbe most often causes respiratory infections?

    a. fungi   b. bacteria   c. viruses

3. Which of the following is not found in soil?

    a. algae   b. protists   c. viruses   d. all of these are found in soil

4. A common characteristic of bacterial pathogens found in the soil is:

    a. spirilla shape   b. cocci shape   c. spore formers   d. Gram negatives

5. The nonliving component of soil is:

    a. humus   b. pumice   c. silicon   d. laterite

WORKING DEFINITIONS AND TERMS

Anthrax A fatal respiratory disease caused by a species of *Bacillus*.

Bacteriophage A virus that infects bacterial cells.

Humus Nonliving organic material found in soil.

Mycoses Fungal infections.

Plaque A clear zone of lysis of bacterial cells on agar, caused by a viral infection.

Primary Pulmonary Infection (PPI) A deep systemic respiratory disease.

Tetanus A neuromuscular disease caused by a species of *Clostridium,* involving sustained muscle contractions.

Topsoil The upper layer of soil, including the surface bacteriophage—a virus that infects bacterial cells.

Weathering The mechanical and physical breakdown of rock, releasing minerals and elements into the soil.

25 Microbial Ecology

Objectives

After completing this lab, you should be able to:

1. Define and understand the concept of ecology.
2. Understand how energy is transferred between organisms.
3. Understand the concept of symbiosis.

4. Understand the symbiotic relationships between humans and microorganisms.
5. Differentiate between microorganisms that are normal flora and those that cause infectious disease.

Ecology is the study of the relationships of organisms and their environment. These relationships include the interactions of organisms with their environment and the interactions of organisms with one another. An *ecosystem* comprises all of the organisms in a given area together with the surrounding physical environment.

The organisms within an ecosystem live in communities. An ecological *community* consists of all the kinds of organisms that are present in a given, specific environment. Communities are made up of *populations,* groups of organisms of the same species. The basic unit of the population is the individual organism. Organisms occupy a particular habitat and niche. The *habitat* is the physical location of the organism. An organism's *niche* is the role it plays in the ecosystem.

Energy is essential to life, and energy from the Sun is the ultimate source of energy for nearly all organisms in the ecosystems on this planet. (The ecosystem of the deep ocean is a noted exception, for sunlight does not penetrate into this environment.) Organisms called *producers (autotrophs)* capture energy from the Sun. They use this energy and various nutrients from soil or water to synthesize the substances they need to grow and support their activities. Energy is transferred when *consumers (heterotrophs)* obtain nutrients by eating the pro-

ducers or other consumers. *Decomposers* obtain energy by digesting dead bodies or wastes of producers and consumers. The decomposers release substances that producers can then use as nutrients. Microorganisms can be producers, consumers, or decomposers in ecosystems.

Microorganisms are important in the recycling of elements, nitrogen, sulfur, phosphorus, and so on, in the physical environment. Because of the vast amount of materials available within this field of microbiology, we will narrow our inquiry to the interrelationships between microorganisms in the ecosystem of the human body. We will also explore the interrelationship between microorganisms and the human organism.

Symbiosis is the interrelationship or association between two or more species. These associations include *mutualism, commensalism,* and *parasitism.* Mutualism is the association in which both members living together benefit from the relationship. For example, *Escherichia coli* live in the human colon. These bacteria provide vitamin K to our body, and they also help us with our digestive processes. The bacteria, in turn, receive nutrients and an environment in which to live. Commensalism is the relationship whereby one organism benefits and the other neither benefits nor is harmed. For example, species of staphylococcus live on the surface of our skin and utilize metabolic products from the pores in our skin. We neither benefit nor are harmed by them, although some suggest that these microorganisms may prevent colonization by other, harmful microorganisms through *microbial competition* for nutrients and space. Parasitism is the relationship whereby one organism, the *parasite,* benefits from the association, whereas the other, the *host,* is harmed or even killed. When a parasitic microorganism invades the human body, it causes an *infectious disease.* Organisms that live on or in the human body but do not cause disease are referred to as *normal microflora.*

Materials List Per Table/Workstation

Melted agar deeps: Trypticase Soy Agar (TSA)

Sterile Petri dishes

Inoculating loops

Waterbath

Penicillium mold culture

Broth cultures of *Serratia marcescens, Proteus mirabilis, S. aureus, Escherichia coli,* colicin strain and noncolicin strain

COMPETITION BETWEEN BACTERIA

PROCEDURE

1. Obtain a tube of melted agar from the waterbath.

2. Quickly, before the agar solidifies, transfer a single loopful of a broth culture of *S. marcescens* into the tube of melted agar. Mix the tube by rolling it between the palms of your hands.

3. Pour the melted agar into a sterile empty Petri dish. (Review Exercise 2 for the pour plate technique.)

4. Replace the dish cover and gently swirl the dish on the table to distribute the agar.

5. Allow the agar to solidify and label the plate. Incubate at 37° C.

6. Repeat the procedure above using a culture of *P. mirabilis* in place of *S. marcescens* in step 2.

7. Repeat the procedure again. However, in step 2, add a loopful of *S. marcescens* and a loopful of *P. mirabilis* to the same melted agar tube.

Results

Count the number of colonies on the pure *S. marcescens* plate. Count the number of colonies on the pure *P. mirabilis* plate.

Count the number of colonies of each organism on the mixed culture plate. The mixed culture plate provides each organism with half of the nutrients and half of the space that they have on the pure culture plates. Comparing its pure plate to the mixed plate, do you see a difference in colony size for each organism?

BACTERIOCIN PRODUCTION

Bacteriocins are proteins that inhibit the growth of other strains of the same species of organism. *E. coli,* is noted for the production of these proteins which are specifically called colicins. Colicins are a different form of competition.

PROCEDURE

1. Repeat the procedure above from steps 1–5, using a culture of *E. coli,* noncolicin strain, in place of *S. marcescens* in step 2.

2. Again, repeat the procedure from steps 1–5, using a culture of *E. coli,* colicin strain, in place of *S. marcescens* in step 2.

3. Repeat the procedure again. However, in step 2, add a loopful of *E. coli,* colicin strain, and a loopful of the noncolicin strain to the same melted agar tube.

Results

Count the number of colonies on the pure *E. coli* colicin plate. Count the number of colonies on the pure noncolicin plate.

Count the number of colonies of each strain on the mixed culture plate. The colicin producing strain acts as a growth inhibitor of the noncolicin strain. Comparing its pure plate to the mixed plate, do you see a difference in colony size of the noncolicin strain?

BACTERIAL-FUNGAL INTERACTION

Bacteria produce acid products when they metabolize. These acid metabolites serve to inhibit the further growth of bacteria by lowering the pH of the environment. Fungi favor a low pH and thus usually replace the presence of bacteria in an environment. More so, the presence of fungi tends to chemically inhibit the growth of bacteria. This is why fungi are common sources of *antibiotics;* antibiotics are compounds that specifically inhibit the growth of bacteria. A common scenario seen in the human body is when a person taking antibiotic medication subsequently suffers a fungal infection.

PROCEDURE

1. Repeat the procedure above from steps 1–5, in "Competition between Bacteria," using a culture of *S. aureus* in place of *S. marcescens* in step 2. Make two plates.

2. Incubate one plate as is.

3. Before incubation of the second plate, using a loop, transfer some *Penicillium* mold to the center surface of this plate. Incubate.

Note: PROCEED WITH CAUTION—Mold spores are easily airborne and can cause infection, allergy, or laboratory contamination.

Results

Count the number of colonies on the pure *S. aureus* plate.

Count the number of colonies of *S. aureus* on the mixed culture plate. Comparing its pure plate to the mixed plate, do you see a difference in colony size for the bacteria?

NAME _____ DATE _____ SECTION _____

QUESTIONS

1. What types of energy roles exist in an ecosystem?

2. What types of symbiotic relationships exist between organisms?

3. What is meant by normal flora?

4. Name some microorganisms that are normal flora and their location in/on the human body.

5. What is the interrelationship between bacteria and fungi?

MATCHING

a. ecosystem

b. habitat

c. consumers

d. symbiosis

e. mutualism

f. host

g. antibiotic

_____ organisms that utilize preformed energy containing substances from producers and/or decomposers

_____ two (or more) organisms within a community where both benefit from the relationship

_____ an organism that is harmed in a parasitic relationship

_____ all organisms within a given area along with their specific environment

_____ any interrelationship between different kinds of organisms

_____ a naturally produced compound that kills or inhibits growth of bacteria

_____ specific physical location of a population

MULTIPLE CHOICE

1. The study of the relationship between organisms and their environment is:
   a. symbiosis   b. niche study   c. ecology   d. competition

2. Two organisms are found within the same environmental area. One benefits from this proximity, whereas the other neither benefits nor is harmed. This relationship is termed:

   a. commensalism   b. autotrophism   c. heterotrophism   d. neutralism

3. Fungi release chemicals that inhibit bacterial growth. This is known as:

   a. parasitism   b. infection   c. commensalism   d. competition

4. Algae are able to absorb carbon dioxide and utilize light to produce glucose. Which term applies:

   a. heterotroph   b. autotroph   c. niche   d. biochemical transference

5. Colicin:

   a. is produced by strains of *E. coli*

   b. is a form of a bacteriocin

   c. inhibits strains of *E. coli*

   d. all of these statements are true

6. Fungi generally prefer:

   a. a low pH

   b. high calcium concentration

   c. low oxygen concentration

   d. deep soil environment

7. Fungi break down such things as leaves and dead animals within the soil. Fungi are therefore:

   a. symbionts

   b. decomposers

   c. commensals

   d. parasites

WORKING DEFINITIONS AND TERMS

Antibiotic   A compound that inhibits the growth of bacteria.

Autotrophs   Organisms that do not require an organic carbon source.

Bacteriocins   Proteins that inhibit the growth of other strains of the same organism.

Commensalism   Relationship whereby one organism in the association benefits and the other does not.

Community   All of the kinds of organisms in a specific environment population; groups of organisms of the same species.

Competition   Aggression for nutrients or space.

Consumers   Organisms that eat the producers or decomposers to obtain energy.

Decomposers   Organisms that digest dead producers and consumers.

Ecology   The study of interrelationships between organisms and their environment.

Ecosystem   All of the organisms in a given area together with their specific environment.

Habitat   The specific physical location of a population.

Heterotrophs   Organisms that must ingest an organic carbon source.

Host   The organism that is harmed in a parasitic association.

Infectious disease   Disease caused by microorganisms.

Mutualism   Relationship whereby both organisms in the association benefit.

Niche   The role of an organism in its ecosystem.

PHOTOGRAPHIC ATLAS

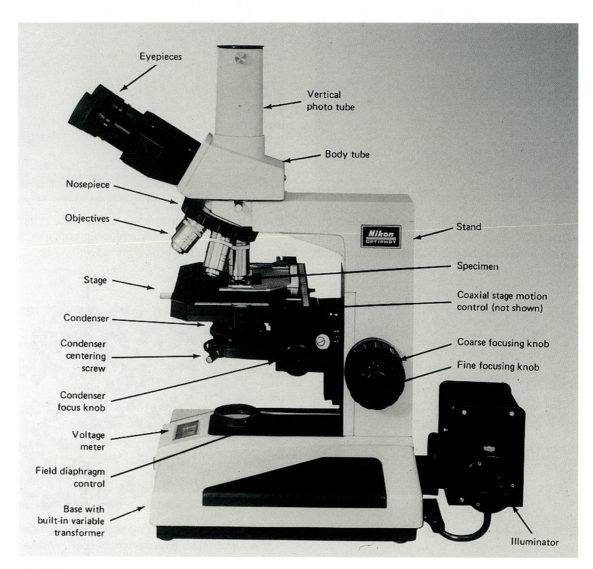

PLATE 1 Compound Microscope

Labels (clockwise from top left):
- Eyepieces
- Vertical photo tube
- Body tube
- Nosepiece
- Objectives
- Stand
- Specimen
- Coaxial stage motion control (not shown)
- Coarse focusing knob
- Fine focusing knob
- Stage
- Condenser
- Condenser centering screw
- Condenser focus knob
- Voltage meter
- Field diaphragm control
- Base with built-in variable transformer
- Illuminator

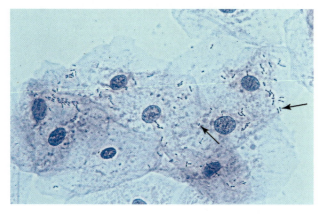

PLATE 2 Simple stain of human cheek epithelial cells. Note that most of these cells have bacteria attached to them (arrows). **LM** 480x

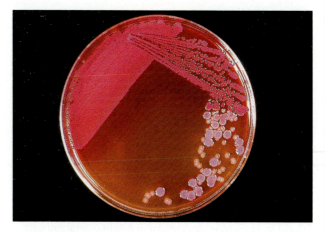

PLATE 3 Streak plate demonstrating the separation, or isolation of two different bacterial types.

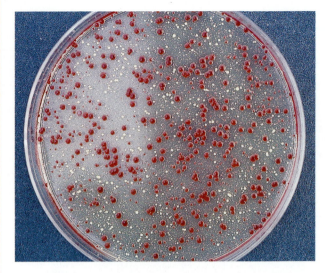

PLATE 4 Pour plate of a mixed culture demonstrating the isolation of the smaller yellow colonies of *Micrococcus luteus* and the larger red colonies of *Serratia marcescens*.

PLATE 5 Bacterial growth characteristics in broth. From left to right: uninoculated tube, precipitation reaction, turbidity, flocculation, pellicle formation.

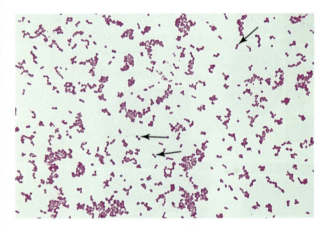

PLATE 6 Simple stain preparation of a diplococcus, *Neisseria gonorrhoeae*. **LM** 743x

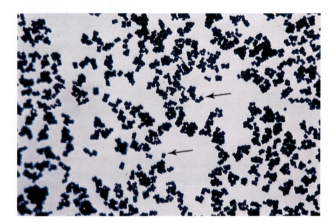

PLATE 7 Simple stain preparation of a sarcinae, or "packet of eight", arrangement of cocci, *Micrococcus luteus*. **LM** 389x

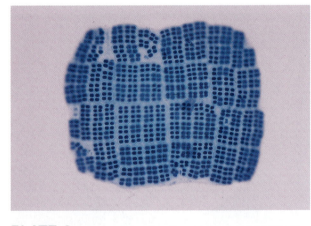

PLATE 8 Simple stain showing a tetrad arrangement of cocci. **LM** 182x

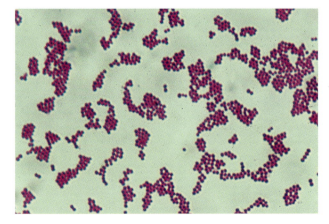

PLATE 9 Simple stain preparation of a staphylococcus, *Staphylococcus aureus*. Note that there is no particular pattern to the grouping of cells. **LM** 1,214x

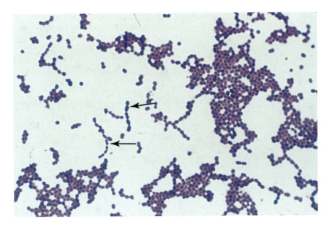

PLATE 10 Simple stain preparation of Enterococcus (formally named *Streptococcus*) faecalis, a streptococcus. Note that within the chains, there is often pairing of the bacteria and when concentrated in larger groups, they appear as staphylococci. **LM** 1,214x

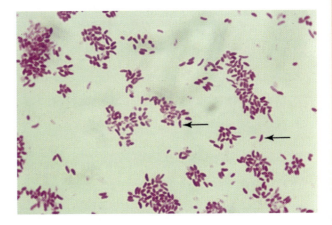

PLATE 11 Simple stain preparation of a vibrio, or "comma" shaped bacillus, *Vibrio cholerae* (formally named *Vibrio comma*), the causative agent of cholera.
LM 1,214x

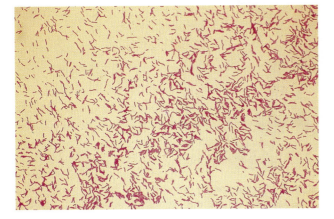

PLATE 12 Simple stain preparation of a typical spirillum **LM** 971x

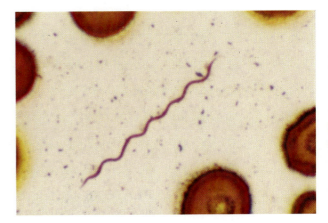

PLATE 13 Simple stain preparation of a spirochete, *Borrelia burgdorferi*, the agent which causes Lyme disease. **LM** 3,750x

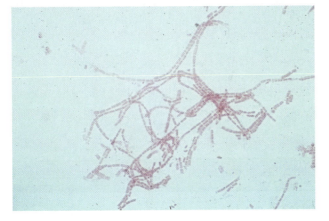

PLATE 14 Simple stain preparation of the streptobacillus, *Bacillus cereus*. **LM** 1,000x

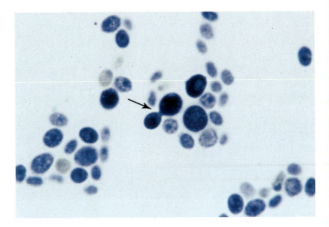

PLATE 15 Simple stain preparation of a yeast, *Saccharomyces cerevisiae*. Note the budding (arrow).
LM 2,500x

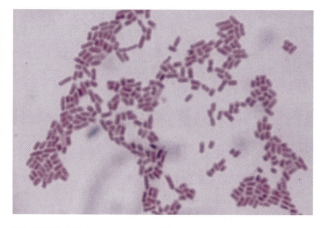

PLATE 16 Gram stain of *Escherichia coli*, showing a negative (safranin colored) reaction. **LM** 1,500x

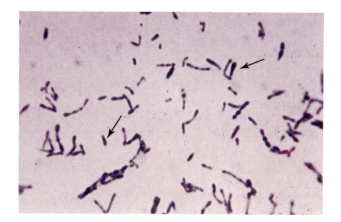

PLATE 17 Gram stain of *Corynebacterium diphtheriae* showing a positive (crystal violet colored) reaction. The arrows point to characteristic club shaped rods. **LM** 2,186x

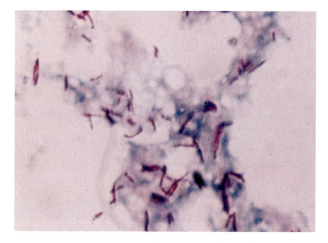

PLATE 18 Acid fast stain reaction of *Mycobacterium tuberculosis* in a sputum sample. The bacteria retains the carbol fuchsin stain while the other materials present are the color of methylene blue or brilliant green. **LM** 2,700x

PLATE 19 Positive spore stain of a bacillus species. The spores retains the malachite green stain while the vegetative cells are the color of safranin. **LM** 2,500x

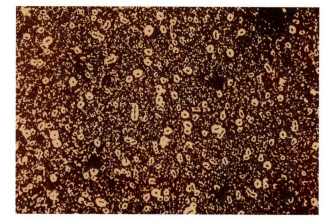

PLATE 20 Capsule stain of *Klebsiella pneumoniae*. The clear areas surrounding the bacterial cells are the capsules. Note that these capsules are thicker than the cells themselves. **LM** 2,250x

PLATE 21 The mold *Rhizopus*, division *Zygomycota*, showing asexual sporangiospore formation within a sporangium. **LM** 250x

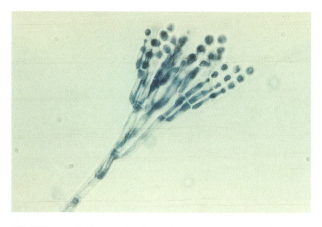

PLATE 22 The mold *Penicillium caseicolum*, division *Ascomycota*, showing conidiospore formation. **LM** 2,286x

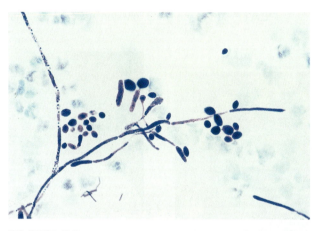

PLATE 23 *Candida albicans*, division *Ascomycota* is a yeast—like fungus. Depending on the environmental temperature, its growth may mimic that of a mold showing pseudohyphae or pseudomycelia. **LM** 625x

PLATE 24 *Rhizopus* thallus completely covering the surface of an agar plate.

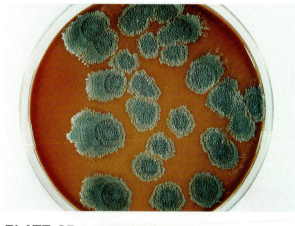

PLATE 25 *Penicillium* mold growing on an agar plate.

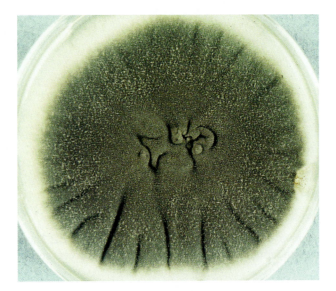

PLATE 26 *Aspergillus*, division *Ascomycota*, thallus.

PLATE 27 Color enhanced image of bacteriophages adsorbing to *Escherichia coli*, its host cell. **LM** 8,500x

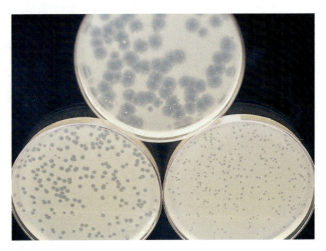

PLATE 28 Plaques of three different *Escherichia coli* specific bacteriophages.

PLATE 29 Colored transmission electron micrograph of a red blood cell infected with the malarial parasite, *Plasmodium*. **LM** 11,000x

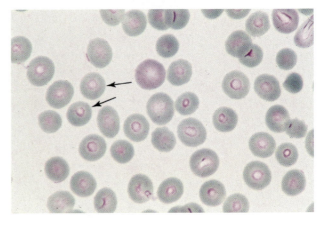

PLATE 30 *Plasmodium* merozoites within human red blood cells. Note the ring formation (arrows). **LM** 1,000x

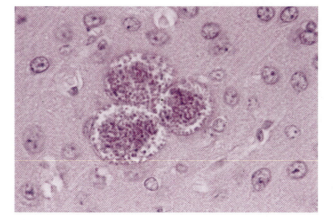

PLATE 31 *Toxoplasma gondii* pseudocysts in brain tissue. **LM** 471x

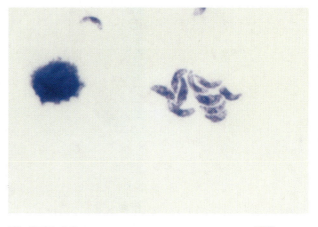

PLATE 32 *Toxoplasma gondii* trophozoites. **LM** 2,500x

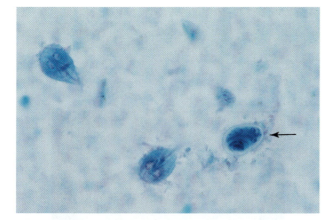

PLATE 33 *Giardia lamblia* trophozoites and cyst (arrow) from human stool sample. **LM** 943x

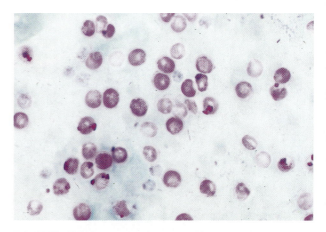

PLATE 34 *Cryptosporidium parvum* in human stool. **LM** 1,000x

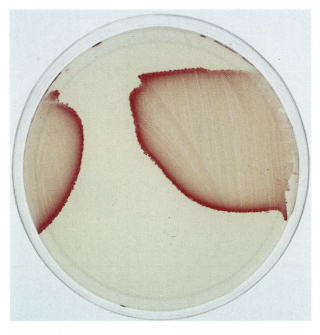

PLATE 35 Demonstration of the poor penetrating power of ultra violet light. Ordinary glasses placed between the light source and an agar plate covered with *Serratia marcescens* prevented the light from reaching the bacteria. The areas of growth correspond to the shape of the lenses of the glasses.

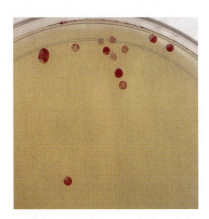

PLATE 36 Demonstration of mutagenic effects of ultra violet light on *Serratia marcescens*. Note the variations of color pigment production among the surviving colonies.

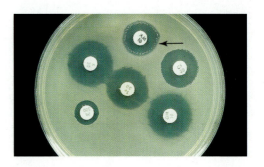

PLATE 37 The Kirby-Bauer antibiotic sensitivity test. The diameter of the zones of inhibition determines whether the drug tested is a candidate for use in a patient. Note the possibility of antibiotic resistance to ampicillin (arrow) as shown by colonies growing just inside the zone of inhibition.

PLATE 39 Oxidation-Fermentation medium with glucose. The microbe inoculated into these tubes is negative for both of these reactions.

PLATE 38 Differential carbohydrate reactions in phenol red broth. The tube on the left demonstrates a negative fermentation reaction, or alkaline conditions. The tube in the center shows fermentation of the carbohydrate, or acid conditions. The tube on the right illustrates fermentation as well as gas formation, or acid plus gas. (Note the gas bubble present in the Durham tube.)

PLATE 40 Oxidation-Fermentation medium with glucose showing oxidation of glucose.

PLATE 41 Oxidation-Fermentation medium with glucose showing fermentation of the carbohydrate.

PLATE 43 Differential reaction of decarboxylase broth. The broth indicates whether the test microbes synthesized decarboxylase, which has the ability to cleave the carboxyl group from certain amino acids. The addition of oil is required to eliminate normal oxidation as a cause for the catabolism of the amino acid. The tube on the left shows a negative reaction whereas the tube on the right is positive.

PLATE 42 Differential reaction of Simmons Citrate medium. The medium in the tube on the left shows a negative reaction. The tube on the right indicates that the bacterium inoculated utilized the citrate for energy production.

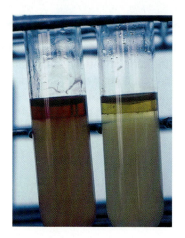

PLATE 44 The Indole test. Kovac's reagent reacts with indole, a product of the catabolism of the amino acid tryptophan. A red color in the tube on the left indicates a positive reaction.

A

B

C

PLATE 45A, 45B, AND 45C Differential results of the Nitrate Reduction test. 46A shows a positive reaction after the addition of reagents A and B (sulfanilic acid and N,N, dimethyl-alpha-naphthylamine.) The red color indicates the inoculated bacterium contains the enzyme to reduce nitrate to nitrite. 46B indicates a negative reaction. After the addition of reagents A and B, no color change was observed. When zinc, a reducing agent, was added to the test medium, the zinc reduced the nitrate to nitrite, triggering the color reaction. 46C shows that the test bacterium had the enzymes to reduce the nitrate to nitrogen gas as indicated by the lack of a color change after the addition of all three reagents.

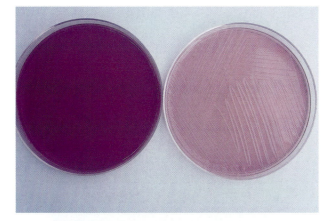

PLATE 46 The Urease test. This reaction is so sensitive that bacterial growth does not have to take place for accurate results to be read or interpreted. The amount of enzyme present in the inoculated bacteria is sufficient to break down the urea to ammonia, resulting in a change of pH. The agar plate on the left demonstrates a positive reaction.

PLATE 47 The Catalase test. When hydrogen peroxide is reacted with the enzyme catalase, it is quickly broken down into water and oxygen, resulting in bubbling. This test can be used to distinguish between catalase positive staphylococci and catalase negative streptococci.

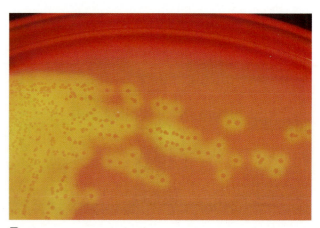

A   B

PLATE 48A AND 48B Demonstration of hemolysis on a blood agar plate.
Alpha (α) hemolytic bacteria, (photo A) partially destroys red blood cells resulting in a light
green zone adjacent to the bacterial growth. Beta (β) hemolytic bacteria, (photo B) completely
destroys the red blood cells resulting in a clear zone adjacent to the growth.

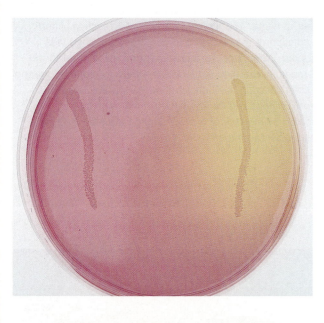

PLATE 49 Reactions of the genus *Staphylococcus* on
Mannitol Salt Agar. This highly selective medium allows only
staphylococci to grow on its surface due to the medium's
high salt concentration. Among the staphylococci, only
S. aureus (on right) is able to ferment the carbohydrate
mannitol, causing a change in pH and thus a color change
of the phenol red indicator from red to yellow. Non mannitol
fermenting *S. epidermidis* is seen growing on the left.

PLATE 50 Reactions of Gram negative bacteria on
MacConkey Agar. Lactose fermenting *Escherichia coli*
(pink colonies) is on the left and non-lactose fermenting
Pseudomonas aeruginosa(colorless colonies) is growing
on the right.

PLATE 51 Reactions on Gram negative bacteria on
EMB agar. Lactose fermenting *Escherichia coli* is on the
left (note the metallic green growth often characteristic of
rapid lactose fermenting bacteria.) Non-lactose fermenting *Pseudomonas aeruginosa* is growing on the right.
The pink colonies are indicative of negative lactose
fermentation.

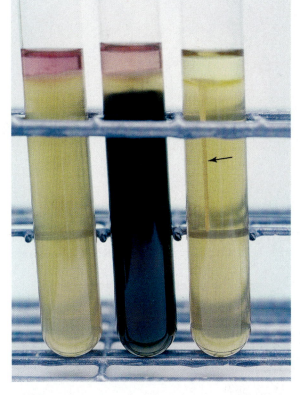

PLATE 53 Reactions in Sulfide-Indole-Motility medium. The tube on the left was inoculated by a bacterium that indole positive as indicated by the color of Kovac's reagent, as well as negative for sulfide production. This microbe is also motile as seen by the cloudiness of the medium. The center tube shows sulfide production as seen by the blackening of the medium. This tube shows indole production as well as motility. The tube on the right shows all negative reactions. Note the sharp line of growth (arrow) indicating that the bacterium inoculated is non-motile.

PLATE 52 Reactions on Triple Sugar Iron Agar. The tube on the left shows all negative reactions. The microbe inoculated failed to ferment any of the carbohydrates, nor did it utilize the iron salt within the medium. The center tube demonstrates glucose fermentation, as seen in the yellow butt, sulfide production, as seen by the blackening of the tube, and negative lactose fermentation as indicated by the red slant. The tube on the right shows glucose and lactose fermentation as seen by the yellow butt and slant, as well as gas formation as seen by the separation of the agar from the tube.

PLATE 54 Indication of a mutation resulting in antibiotic resistance as seen by colonial growth well within the zone of inhibition of a Kirby-Bauer antibiotic sensitivity plate. (Antibiotic disk on the upper left.)

PLATE 55 The Ames test. Note that the mutagenic agent within the disk triggered a back mutation that allowed the test microbe to synthesize histidine, and thus grow.

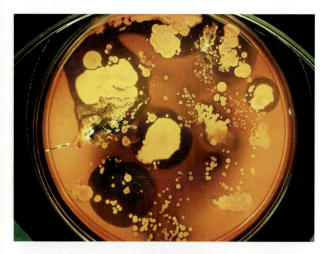

PLATE 56 Blood Agar Plate exposed to unwashed hands. Note the variety of different morphological types.

PLATE 57 Bacitracin sensitivity test for *Streptococcus pyogenes.*

PLATE 58 The CAMP test for identification of *Streptococcus agalactiae.* The overlapping of the diffused hemolytic enzymes from *S. agalactiae* and *Staphylococcus aureus* results in an extremely clear zone of beta hemolysis, indicative of a positive reaction. The addition of a bacitracin disk within the inoculated *S. agalactiae* serves as a check for *Streptococcus pyogenes.*

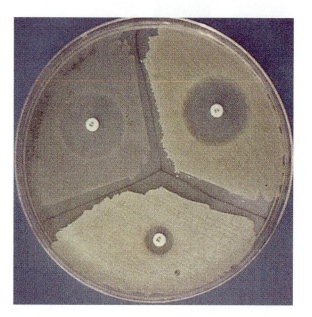

PLATE 59 Novobiocin sensitivity. *Staphylococcus saprophyticus* (bottom) is resistant to this antibiotic which differentiates it from other species within this genus.

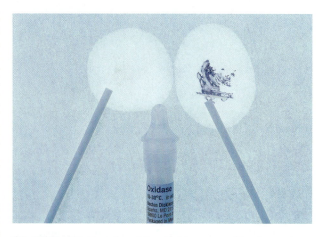

PLATE 60 Results of the Oxidase test. The bacterium on the right is positive for this enzyme whereas the one on the left is negative.

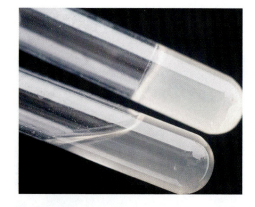

PLATE 61 Coagulase Test. The tube on the top shows a positive coagulase reaction in rabbit plasma typical of pathogenic *Staphylococcus aureus.* The tube on the bottom has been inoculated with coagulase negative *Staphylococcus epidermidis.*

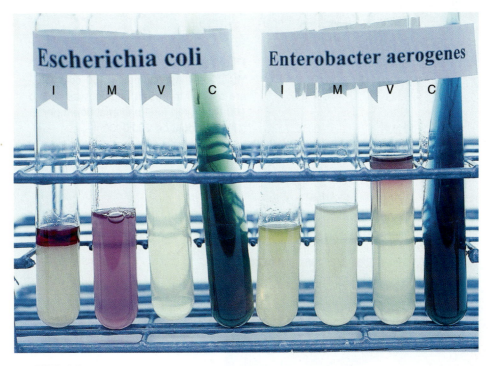

PLATE 62 The IMViC (Indole, Methyl Red, Voges Proskauer, Citrate) set of reactions used to differentiate between *Escherichia coli* and *Enterobacter aerogenes*. *E. coli* demonstates a typical (+),(+), (–),(–) pattern of reactions and *E. aerogenes* gives the opposite (–), (–),(+),(+) reactions.

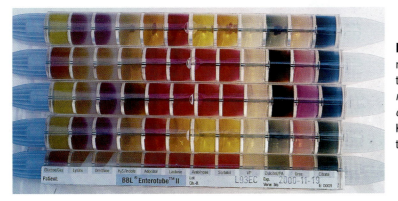

PLATE 63 Reactions in the Enterotube® II rapid identification system. Shown growing from top to bottom: *Serratia marcescens, Pasturella multocida, Proteus mirabilis*, and *Escherichia coli*. The bottom tube is an uninoculated control. Kovac's reagent was not added to any of the tubes.

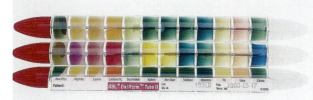

PLATE 64 Reactions of the Oxi/Ferm™ Tube II. The top tube shows reactions of a *Flavobacterium spp*. The center tube was inoculated with *Pseudomonas aeruginosa*. The bottom tube in an uninoculated control. Kovac's reagent was not added to any of the tubes.

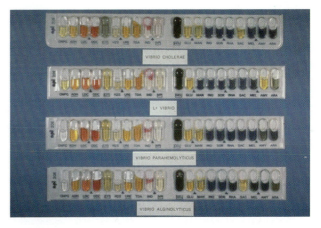

PLATE 65 Reactions in the Analytical Profile Index (API) 20E.

PLATE 66 Colony count on Tryptone Yeast Agar. Approximately 200 colonies are seen. If the water sample was diluted 100 times, the total bacterial count per milliliter would be 200,000/ml.

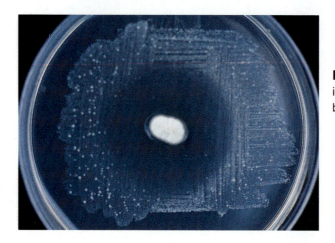

PLATE 67 *Penicillium notatum* inhibiting the growth of competitive bacteria.

Normal flora   Microorganisms that inhabit the human body without doing harm.

Parasite   The organism that benefits in a parasitic association.

Parasitism   Relationship whereby one organism in the association benefits and the other is harmed.

Producers   Autotrophs that capture the energy of the Sun.

Symbiosis   The interrelationship/association between organisms of different species.

Bibliography

Alcamo, IE, *Fundamentals of Microbiology,* Jones and Barlett Publishers, 2000.

Black, J. *Microbiology, Principles and Exploration,* 4th Edition, Prentice-Hall, Upper Saddle River, N.J. 1999.

Cappuccino, J, Sherman, N. *Microbiology, A Laboratory Manual,* 5th Edition, Addison Wesley Longman Inc., Menlo Park, Ca. 1999.

Clark, George. Editor. *Staining Procedures,* 4e, Williams & Wilkens, Baltimore/London, 1981.

De Kruif, P, *Microbe Hunters,* Harcourt Brace & Company, New York, NY, 1926.

DIFCO Manual, 9th Edition, DIFCO Laboratories Inc. Detroit, Michigan, 1969.

DIFCO Manual, 10th Edition, DIFCO Laboratories Inc. Detroit, Michigan, 1984.

Dirckx, J, *Stedman's Concise Medical Dictionary For The Health Professions,* 3rd Edition, Williams & Wilkins, Baltimore, MD, 1997.

Forbes, B, Sahn, D, Weissfeld, A, *Baily & Scott's Diagnostic Microbiology,* 10th Edition, Mosby Inc. St. Louis, MO, 1998.

Gladwin, M, Trattler, B, *Clinical Microbiology Made Ridiculously Simple,* 2nd Edition, Medmasters Inc. Miami, Fla. 1997.

Howard, B, Klass, J, Rubin, S, Weissfeld, A, Tilton, R. *Clinical and Pathogenic Microbiology,* C.V. Mosby Company, St. Louis, MO, 1987.

Johnson, T, Case, C, Laboratory Experiments in Microbiology, 6th Edition, Benjamin Cummings, San Francisco, Ca. 2001.

Koneman, E, Allen, S, Dowell, V.R, Sommers, H, *Color Atlas and Textbook of Diagnostic Microbiology,* 2nd Edition, J.B. Lippincott Co, New York, NY, 1983.

Lennette, E, Spaulding, E, Truant, J. *Manual of Clinical Microbiology,* 2nd Edition, American Society for Microbiology, 1974.

Payne-Palacio, J, Theis, M, Introduction to Foodservice, 9th Edition, Prentice-Hall Publishers, Upper Saddle River, NJ, 2001.

Pierce, B, Leboffe, M, *Exercises for the Microbiology Laboratory,* Morton Publishing Co, Englewood, CO, 1999.

Schaechter, M, Neidhardt, F, Ingraham, J, *An Electric Companion To Beginning Microbiology,* Cogito Learning Media Inc, New York, NY, 1997.

Shimeld, L, Essentials of Diagnostic Microbiology, Delmar Publishers, 1999.

Wistreich, G, *Microbiology Perspectives,* Prentice-Hall, Upper Saddle River, NJ. 1999.

1 Flow Chart for the Identification of Enterobacteriaceae

Microbes: Gram (−), oxidase (−), glucose fermentation (+)

GROUP I

REACTION

Phenylalamine deaminase (PD or PAD) positive
  or
Tryptopane Deaminase (TDA) positive

POSSIBLE ORGANISMS

Proteus or *Providencia*

Urease (−) *Providencia stuartii* or *alcalifaciens*
    Do inositol test (carbohydrate)
      Inositol (+) *Providencia stuartii*
      Inositol (−) *Providencia alcalifaciens*

USE THE FOLLOWING CHART

	ODC (ORNITTHINE DECARBOXYLASE)	H$_2$S	INDOLE	RHAMNOSE
Pr. mirabilis	+	+	−	
Pr. vulgaris	−	+	+	
M. morganii	+	−	+	
Prov. rettgeri	−	−	+	+
Prov. stuartii	−	−	+	−

REACTION

PD or TDA negative
Voges Proskauer positive

POSSIBLE ORGANISMS

Klebsiella, Enterobacter, Serratia, Hafnia
(Note: Some species of *Klebsiella* are
VP (−). These can be found in Groups V and VI)

Gelatin Positive   *Serratia*

USE THE FOLLOWING CHART

	ODC ORNITTHINE DECARBOXYLASE	SOR (SORBITOL)	ARA (ARABINOSLE)	RED PIGMENT*	INDOLE	SUCROSE
S. marcescens	+	+	−	+	−	+
S. liquifaciens	+	+	+	−	−	+
S. rubidaea	−	−	+	+	−	+
S. odorifera (type I)	+	+	+	−	+	+
S. odorifera (type II)	−	+	+	−	+	−

*This is not a reliable test as the pigment does not show up under different growth conditions

Gelatin Negative   *Klebsiella* and *Enterobacter*

LOOK AT DECARBOXYLASE REACTIONS

Lysine positive   indole negative—   *Klebsiella pneumoniae.*   All carbohydrates should be positive.
                   indole positive—   *Klebsiella oxcytoca.*   All carbohydrates should be positive.
Lysine, ornithine positive—   *Enterobacter aerogenes.*   All carbohydrates should be positive
                               Enterobacter gergoviae.   Urease will be positive, sorbitol will be negative
                               Enterobacter hafnia   (May be listed as a separate genus). Glucose, manittol, rhamnose, Amylose and/or arabinose positive. Citrate negative.
Arginine, ornithine positive—   *Enterobacter cloaceae*   Sorbitol positive, inositol negative
                                 Enterobacter sakazakii   Sorbitol negative, inositol usually positive
Ornithine positive—   *Enterobacter gergoviae*   Urease positive, sorbitol negative
All decarboxylase reactions negative—   *Enterobacter agglomerans.*   Variable carbohydrate reactions.

GROUP III

REACTION

PD or TDA negative
Voges Proskauer negative
Hydrogen sulfide positive

POSSIBLE ORGANISMS

Salmonella, Arizona, Citrobacter freundii
Edwardsella tarda (some E. coli)

Lysine decarboxylase positive

USE THE FOLLOWING CHART

	ONPG	INOSITOL	INDOLE
Salmonella	–	+	–
Arizona	+	–	–
Edwardsella tarda	–	–	+
E. coli (rare)	+	–	+
Salmonella typhi	–	–	–

Lysine decarboyxlase negative
    Citrobacter freundii—amylase may be positive or negative
                Indole negative

GROUP IV

REACTION

PD or TDA negative
Voges Proskauer negative
Hydrogen sulfide negative
Indole positive

POSSIBLE ORGANISMS

E. coli, Shigella, Yersinia, Citrobacter diversus

LOOK AT DECARBOXYLASE REACTIONS

Lysine positive—   *E. coli*   carbohydrates variable, citrate negative
Lysine negative—   *Shigella* species. See chart or do serology
                  Citrobacter diversus   ONPG positive, Amygdalin positive (35%), Saccharose positive (50%)
                  Yersinia enteroclitica   urea positive

GROUP V

REACTION

PD or TDA negative
Voges Proskauer negative
Hydrogen sulfide negative
Indole negative
ONPG positive

POSSIBLE ORGANISMS

Shigella sonnie, Klebsiella, Yersinia Hafnia
(new classification of *Enterobacter hafnia*),
Serratia, Enterobacter

LOOK AT DECARBOXYLASE REACTIONS

Lysine and Ornithine positive
Hafnia	glucose, mannitol, amygdalin and/or arabinose positive Inositol, saccharose, melibiose negative
Serratia marcescens	Rhamnose, arabinose negative citrate—80% positive
Serratia liquifaciens	arabinose, citrate positive
Serratia fonticola	inositol, sorbitol, melibiose, amygdalin positive saccharose negative
Serratia odorifora	citrate, and all sugars positive

Arginine and Ornithine positive
Enterobacter sakazakii	inositol, meliobiose, saccharose, amygdalin positive sorbitol negative

Ornithine positive
Shigella sonnei	mannitol, rhamnose, arabinose positive

Lysine positive
Serratia odorifera	citrate, saccharose positive All other sugars negative

All Decarboxylase reactions negative

USE THE FOLLOWING CHART

	UREASE	SORBITOL	RHAMNOSE	MELIOBIOSE
Yersinia pestis	–	+	–	–
Klebsiella ozaenae	–	VAR	VAR	+
Yersinia pseudotuberculosis	+	–	+	–
Enterobacter agglomerans	+	–	+	–
Serratia plymuthica	–	–	–	+

Flow Chart for the Identification of Enterobacteriaceae

GROUP VI

REACTION

PD or TDA negative
Voges Proskauer negative
Hydrogen sulfide negative
Indole negative
ONPG positive

POSSIBLE ORGANISMS

Klebsiella, Salmonella

LOOK AT THE DECARBOXYLASE REACTIONS

Lysine positive —

Hafnia (Enterobacter hafnia)

Salmonella typhi   mannitol, sorbitol, meliobiose positive

Ornithine positive —

Salmonella paratyphi

All Decarboxylase reactions negative   *Klebsiella rhinoscleromatis*

Gram (−) Flow Chart

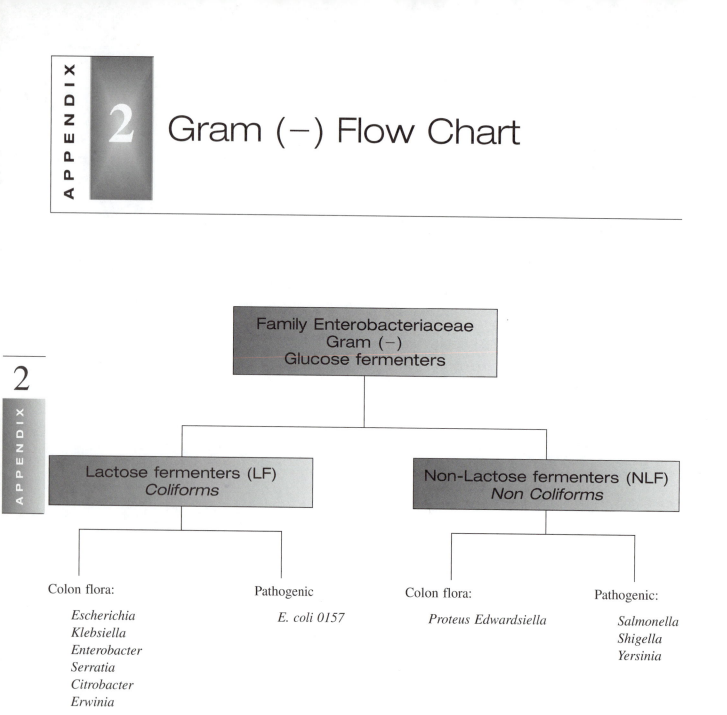

Family Enterobacteriaceae
Gram (−)
Glucose fermenters

Lactose fermenters (LF)
Coliforms

Non-Lactose fermenters (NLF)
Non Coliforms

Colon flora:

Escherichia
Klebsiella
Enterobacter
Serratia
Citrobacter
Erwinia

Pathogenic

E. coli 0157

Colon flora:

Proteus Edwardsiella

Pathogenic:

Salmonella
Shigella
Yersinia

Gram (+) Flow Chart

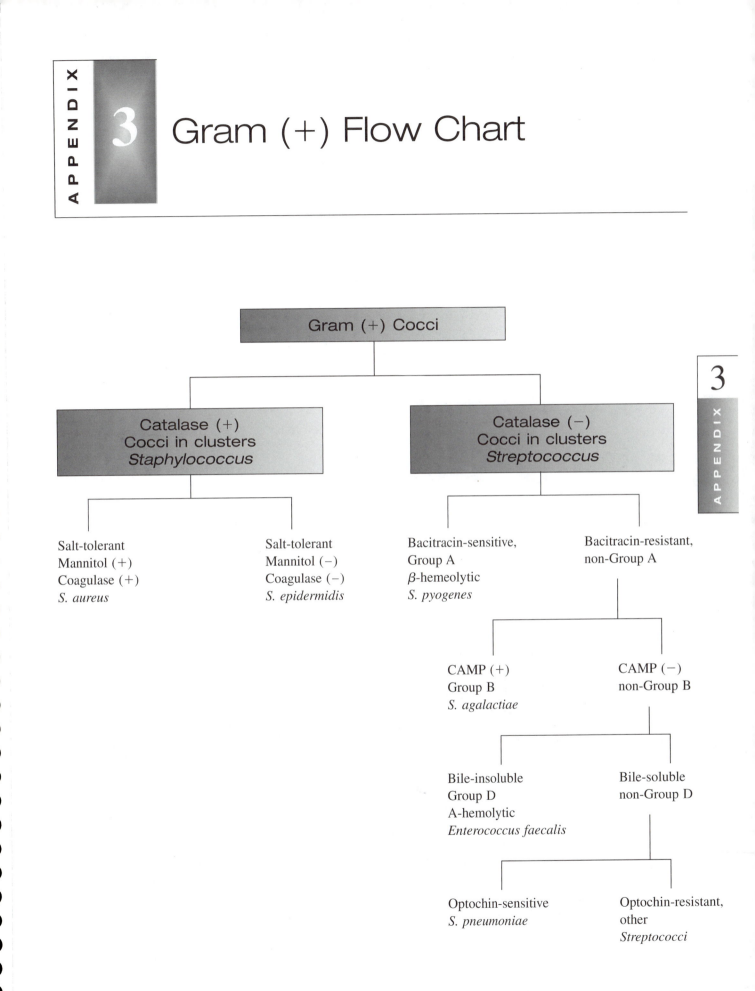

Gram (+) Cocci

Catalase (+)
Cocci in clusters
Staphylococcus

Catalase (−)
Cocci in clusters
Streptococcus

Salt-tolerant
Mannitol (+)
Coagulase (+)
S. aureus

Salt-tolerant
Mannitol (−)
Coagulase (−)
S. epidermidis

Bacitracin-sensitive,
Group A
β-hemeolytic
S. pyogenes

Bacitracin-resistant,
non-Group A

CAMP (+)
Group B
S. agalactiae

CAMP (−)
non-Group B

Bile-insoluble
Group D
A-hemolytic
Enterococcus faecalis

Bile-soluble
non-Group D

Optochin-sensitive
S. pneumoniae

Optochin-resistant,
other
Streptococci

Index

Pure culture, 19
Pyruvate, 175

Q

Quantitation of microbial number in water sample, 195, 196
Quantitative urinalysis, 141–144

R

Rapid slide test for coagulase, 151
Reducing agents, 104
Reservoir of infection, 133
Resistant, 93
Retinitis, 76
Rheostat, 8
Rhizopus, 64
Ringlike trophozoite, 77
Rinsing off a slide, 5
Rod, 33, 34

S

Saccharomyces, 65
Saccharomyces cerevisiae, 64
Safety rules, i, 3, 4, 10
Safranin, 55, 57
Salmonella, 199–201
Salmonella typhimurium, 127
Sarcinae, 33, 34
Selective media, 117, 118
Sensitive, 93
Septa, 63
Septate hyphae, 63, 64
Serial dilution, 71
Serology, 155–163
    ELISA test, 157, 158
    febrile agglutinins test, 156, 157
    latex agglutination, 158–160
Serratia marcescens, 127
SIM, 118
Simmons citrate medium, 103
Simple stain, 4
Simple stain procedure, 35
Simple stain technique, 4
Single rod, 34
Slant-to-broth transfer, 18
Slant-to-slant transfer, 18
Slidex Strepto-Kit, 159
Slow freezing, 91
Smear preparation, 33
Snapping, 34
Soil microbiology, 205–210
Specialized media, 117–124
Specificity, 155

Specimen-handling protocols, 139–148
    general principles, 139
    GI tract, 144
    quantitative urinalysis, 141–144
    throat cultures, 140, 141
Spiral, 34
Spirillum, 33, 34
Spirochete, 34
Spontaneous mutation, 127
Sporangia, 64
Spore coat, 55
Spore stain technique, 56
Sporozoa, 75
Sporozoites, 76, 77
Stain
    acid-fast, 48
    capsule, 57, 58
    endospore, 55, 56
    Gram, 47–49
    primary, 47
Standard plate count method, 195
Staphylococcus, 33, 34
Staphylococcus aureus, 40, 118, 149, 150
Stationary phase, 40
Streak plate, 19–21
Streptobacilli, 34
Streptococci differentiation using latex agglutination, 158–160
Streptococcus, 33, 34
Streptococcus agalacitae, 150
Streptococcus pneumoniae, 56
Streptococcus pyogenes, 112, 140, 150, 157, 158
Strict anaerobes, 111
Substrates, 101
Sulfide-indole-motility (SIM), 118
Sulfur, 41
Superoxides, 111
Susceptible, 93
Symbiosis, 75, 211

T

T-Soy agar, 166
Tachyzoite, 78
TDT, 90
10X objective, 6
Tetanus, 205
Tetrad, 33, 34
Thermal death time (TDT), 90
Thioglycollate, 112
Throat culture procedure, 119
Throat cultures, 140, 141
Throat swab procedure, 141
Too numerous to count (TNTC), 142